The CHEMISTRY of GERMANIUM

The CHEMISTRY
of GERMANIUM

FRANK GLOCKLING

Professor of Inorganic Chemistry
The Queen's University, Belfast, Northern Ireland

 1969

ACADEMIC PRESS · LONDON & NEW YORK

ACADEMIC PRESS INC. (LONDON) LTD.
Berkeley Square House
Berkeley Square
London, W.1

U.S. Edition published by
ACADEMIC PRESS INC.
111 Fifth Avenue
New York, New York 10003

Library of Congress Catalog Card Number: 68–57924

Printed in Great Britain by
Spottiswoode, Ballantyne & Co. Ltd., London and Colchester

PREFACE

This monograph is designed partly to appeal to the general chemical reader who is likely to be interested in the chemistry of germanium, especially in relation to silicon, tin and lead. Sufficient detail has also been incorporated in an attempt to provide a useful source of reference for specialist research workers.

The primary literature has been covered up to early 1968, although Russian publications, examined in English translation, are less up-to-date.

The author is greatly indebted to his wife for extensive work on checking both text and references.

Durham   F. GLOCKLING
September, 1968

CONTENTS

INTRODUCTION

The existence of germanium was first suspected by Newlands (1864) as the missing member of the triad silicon, germanium, tin. Mendeleeff (1871) recognized it as a missing element in his periodic table and predicted its general chemical properties from its relationship to silicon and tin. It was first isolated as the sulphide GeS_2 from the rare mineral argyrodite, $4AgS.GeS_2$ (Winkler, 1887). The development of its chemistry has been slow compared to its neighbours; this may be ascribed to its comparative rarity and high price, and because early studies established that its properties seemed to fit so clearly between those of silicon and tin, that its chemistry appeared to lack novelty. Germanium metal, the dioxide and tetrahalides are now commercially available in quantity, and publications on inorganic and organic compounds of germanium are appearing at an ever increasing rate. Despite this upsurge in research, the only commercial outlets are based on the semi-conductor properties of the metal, the use of germanium dioxide in the production of optical glass of high refractive index and the addition of germanium or its compounds to lead–acid accumulators to reduce the cell resistance.

The important fungicidal properties of organostannanes has led to similar studies on germanium compounds, but without revealing any added specificity or extending the range of antimicrobial activity (Sijpesteijn *et al.*, 1964). The emergence of new chemical applications for germanium compounds will probably depend on the discovery of highly specific, possibly catalytic, processes; otherwise what can be done with germanium can almost certainly be achieved more cheaply with silicon or tin analogues.

REVIEW ARTICLES

Inorganic compounds of germanium were reviewed by Johnson (1952), and considerable advances in the subject have been made since that date. Organogermanium chemistry has been the subject of several reviews (Johnson, 1951; Rijkens, 1960; Quane and Bottei, 1963; Glockling, 1966). An important compilation of data, largely in tabular form (Dub, 1967), is complementary to this book, which does not include lists of compounds.

1 | BOND PROPERTIES

I. ELECTRONEGATIVITY

The electronegativity of germanium relative to the other group IV elements has been the subject of several papers and a great deal of confusion. Electronegativity may be defined as the power of an atom in a molecule to attract electrons to itself, and Pritchard and Skinner (1954), in an excellent review, point out that because electronegativity is concerned with atoms in molecules rather than atoms in isolation, it is not possible to define precise electronegativity values. The electronegativity of an element is influenced by its molecular environment in such a variety of ways as to preclude exact measurement. The scales of electronegativity that have been formulated are discussed by various other authors (Allred and Rochow, 1958; Drago, 1960; Allred, 1961) and these raise a contentious point in connection with the group IV elements carbon, silicon, germanium, tin, lead. All scales are in agreement that carbon is more electronegative than the other elements; Pauling's scale and its modified versions propose a progressive decrease with increasing atomic number (C ~ 2·5; Si, 1·8; Ge, 1·7; Sn, 1·7).

Allred and Rochow have pointed out that in some of its reactions germanium appears to be more electronegative than silicon or tin. However the apparent anomalies are found in mechanistically complex reactions, some of which, such as the reduction of organogermanium halides by zinc and hydrochloric acid, proceed in low yield. Further illustrations that appear to support this view are the relative acid strengths of β-carboxy-silanes and -germanes ($Me_3SiCH_2CO_2H$, $K_a = 0·6 \times 10^{-5}$; $Me_3GeCH_2CO_2H$, $K_a = 1·59 \times 10^{-5}$), and the direction of hydrogen chloride cleavage in mixed silyl–germyl compounds (Mironov and Kravchenko, 1963)—

$$Me_3SiCH_2HgCH_2GeMe_3 \xrightarrow{\text{HCl}} Me_4Ge + Me_3SiCH_2HgCl$$

The "alternating" electronegativity scale produced by Allred and Rochow from thermochemical data (C, 2·60; Si, 1·90; Ge, 2·00; Sn, 1·93; Pb, 2·54) has been severely criticized by Drago (1960), who considers that the electronegativities of silicon to lead are either essentially constant or decrease slightly as in the original Pauling scale. Relative proton magnetic resonance (p.m.r.) shifts in organo-silanes, -germanes and -stannanes have been invoked in support of the "alternating" scale, but again the interpretation is suspect. Other physicochemical measurements discussed later in this

3

Chapter, such as relative reaction rates, favour a progressive decrease in electronegativity with atomic number. Probably the controversy is the result of trying to attach too quantitative a meaning to electronegativity.

II. BONDING IN GERMANIUM COMPOUNDS

Like electronegativity, the concept of π-bonding involving unoccupied $4d$ orbitals of germanium has passed through a controversial stage. There is scarcely an anomaly in germanium chemistry that has not been ascribed to π-bonding; it is only in the past few years that more definitive experimental work has been reported.

In most of its compounds germanium forms four sp^3 tetrahedrally disposed σ-bonds to other elements, and there is no evidence for p_π-p_π multiple bonding as in carbon compounds. When bonded to electronegative elements, the co-ordination number can increase to 6 as in $(Me_3N)_2GeF_4$. These are octahedral d^2sp^3 complexes in which all of the bonds to germanium have σ-symmetry. In its divalent state, germanium has not yielded stable monomeric compounds, and the dihalides, such as GeF_2, are associated by bridging fluorine groups. Monomeric alkyls and aryls, R_2Ge, formally analogous to carbenes, probably exist as reaction intermediates, and they are certainly produced by electron-impact-induced fragmentation processes.

In compounds, where germanium is directly bonded to an aromatic ring, an olefinic or acetylenic group or an element containing doubly occupied non-bonding p or d orbitals, there exists the possibility of partial π-bonding to the unoccupied $4d$ orbitals of germanium. On the basis of high metal–carbon overlap integrals for the elements silicon, germanium, tin and lead Craig *et al.* (1954) concluded that d_π-p_π bonding may be important in their derivatives with unsaturated carbon compounds of the types M—aryl, M—CH=CR$_2$ and M—C≡CH. These calculations also suggest that overlap is not critically dependent on the size of the d orbital. The extent of π-bonding also depends on the relative energies of the participating orbitals, and all available evidence suggests that these two factors are most favourable with silicon, and decrease progressively with increasing atomic number. The extent to which the lone pair of electrons on nitrogen is delocalized into the d orbitals of silicon, germanium and tin in the amines $(Me_3M)_3N$ has been examined by Perkins (1967) using self-consistent field calculations. The results are consistent with extensive d_π-p_π bonding, and lead to an average silicon—nitrogen π-bond energy of about 16 kcal. mole^{-1}. Non-planarity in molecules of this type will reduce, but not necessarily destroy, π-bonding.

Many physicochemical studies indicate that in neutral molecules the effects of π-bonding can be detected and are most pronounced with silicon.

Factors such as the relative base strengths of silyl and germyl oxides and amines and the acid and base strengths of silanols, germanols and stannols, are considered in later Chapters. In silicon chemistry, silicon—oxygen and silicon—chlorine bond lengths have been invoked as evidence of π-bonding, e.g., H_3Si—Cl, 1·28 Å and CH_3—Cl, 1·87 Å (Cruickshank, 1961), as also has the small low-field shift of the methyl proton resonance in $(Me_3Si)_2O$ and $(Me_3Si)_3N$. Dipole-moment measurements on phenyl derivatives of the group IV elements are also indicative of π-bonding, especially with phenyl-silanes (Soffer and De Vries, 1951; Huang and Hui, 1964). For the phenyl-trimethyl derivatives, Me_3MPh, the values are: Me_3CPh, 0·55; Me_3SiPh, 0·44; Me_3GePh, 0·38; and Me_3SnPh, 0·51 D. In the absence of π-bonding, dipole moments should increase with decreasing electronegativity of the metal. Thus the inductive (electron release, $+I$) effect is greatest for tin, whereas the π-bonding (electron attracting, $-M$) effect is greatest with silicon, germanium occupying an intermediate position. The dipole moments of silyl- and germyl-halides, obtained from their microwave spectra, are in agreement with a greater degree of π-bonding between silicon and halogen atoms (H_3SiCl, 1·303; H_3GeCl, 2·124 D) (Mays and Dailey, 1952).

Relative π-bonding effects have been investigated by measuring the rate of removal of the alkynyl proton by hydroxide ion in two series of tritium-labelled compounds, $Et_3MC\colon C^3H$ and $Et_3MCH_2C\colon C^3H$. (Eaborn et al., 1966b). The transition state in these reactions is believed to lie close to the carbanion $RC\colon C^-$ and water; hence electron-withdrawing groups will increase the rate. Pseudo first-order rate constants are shown in Table I.

TABLE I

Rates of de-tritiation of $XC\colon C^3H$ **at pH 8·05 and 25°C**

X	$10^4\,k$, min^{-1}	X	$10^4\,k$, min^{-1}
Me_3Si	65·9	Me_3CCH_2	6·97
Et_3Si	35·7	Me_3SiCH_2	3·58
Et_3Ge	31·9	Et_3SiCH_2	2·40
Me_3C	8·93	Et_3GeCH_2	1·65

For the compounds having a methylene group adjacent to the triple bond, the changes in rate are determined largely, if not entirely, by inductive effects. The $+I$ effect of trialkyl-silyl and -germyl groups is greater than that of the t-butyl group, whereas the difference in rate between the Et_3SiCH_2 and Et_3GeCH_2 groups is small, but consistent with a lower electron release from silicon than germanium. The relative rates for the compounds in which the group IV element is adjacent to the triple bond can be attributed to π-bonding effects more than offsetting inductive effects. Thus for silicon and germanium in the compounds $R_3MC:C^3H$, the electron-withdrawing effect of π-bonding is considered to stabilize the negatively charged transition state more than the initial state.

A further approach to the study of π-bonding comes from an examination of the electronic absorption spectra of phenyltrimethyl derivatives of the group IV elements (Musker and Savitsky, 1967). The spectroscopic moment of phenyl substituents can be measured by an analysis of the intensity change of the symmetry forbidden $A_{1g} \rightarrow B_{2a}$ transition, near 2600 Å; the sign of the moment depending on the electronic effect of the substituent. Thus, bromo-, iodo- and methoxy-benzene have positive moments, and hence these substituents donate electrons. In p-bromo- and o-iodo-anisole, the oscillator strength is smaller than that of anisole indicating that bromo- and iodo-groups behave as electron-withdrawing groups when there is a strong donor group in the *para* position. This inversion in the sign of the spectroscopic moment of the halogens has been attributed to π-bonding between unoccupied d orbitals on the halogen and π-electrons of the ring. Of the halogens, only fluorine retains its positive spectroscopic moment with a strong donor group in the *para* position. Application of this method to trimethyl(4-methoxyphenyl)–metal compounds shows that Me_3M groups either produce an inversion from electron-donating to electron-accepting effect or decrease the spectroscopic moment to zero. The ratio of observed oscillator strengths to that of anisole give the values for Me_3M of 1·00 when M = C, 0·80 when M = Si, 0·86 when M = Ge and 1·00 when M = Sn, and these values have been taken as a measure of the extent of overlap between metal d orbitals and the e_{1g} orbitals of the benzene ring. In similar trimethylsilyl- and trimethylgermyl-biphenyls, Me_3M—C_6H_4Ph, Curtis *et al.* (1967) find that the band near 2500 Å ($\pi_m \rightarrow \pi_{m+1}$) moves to longer wavelengths between carbon and silicon (Me_3C, 257; Me_3Si, 262; Me_3Ge, 260; Me_3Sn, 262 nm). Hückel treatment of the silicon and germanium compounds shows that the Me_3M group provides a high-energy vacant orbital that withdraws electron density from the aromatic system and lowers the energies of π_m and π_{m+1}. The effect on the energy of π_{m+1} is the greater so that the absorption band shifts to longer wavelengths relative to biphenyl.

Electron spin resonance (e.s.r.) spectra of anions of phenyl-silanes and -germanes provide a measure of the extent to which an electron in the e_{2g} orbital of benzene is delocalized over the d orbitals of the metal. (Bedford *et al.*, 1963; Curtis and Allred, 1965). Aryl derivatives of the group IV metals react with alkali metals in solvents such as tetrahydrofuran to form paramagnetic charge-transfer complexes that are stable only at low temperature. Decomposition is rapid at 20°C, producing diamagnetic products—

$$Me_3GePh + K \xrightarrow[-50°C]{THF} [Me_3GePh]^{-•} K^+$$

In benzene, the six π electrons occupy the three lowest-energy molecular orbitals, $(a_1, e_1^{\pm})$ (Fig. 1), and in the benzene anion $(C_6H_6)^{-•}$ the additional electron is considered to be equally distributed between the degenerate antibonding orbitals $(e_2^{\pm})$. When a substituent is introduced into the benzene ring, the degeneracy of the e_2 orbital is removed, and in odd-electron ions such as $(Me_3GePh)^{-•}$, the additional electron will occupy either the *symmetric* or the *antisymmetric* orbital, whichever has the lower energy. Bolton and Carrington (1961) have shown that if a substituent is electron repelling, like the t-butyl group, the electron occupies the *antisymmetric* orbital, and its density at the substituent position is zero. Conversely, for substituents that are electron attracting, the added electron occupies the *symmetric* orbital. The electron distributions in the two cases are shown in Fig. 1.

With the anions of trimethylphenyl-silane and -germane, e.s.r. spectra show that the added electron occupies the *symmetric* orbital; hence both Me_3Si and Me_3Ge groups are electron attracting. From the values of the aromatic proton and methyl proton hyperfine splitting constants it is apparent that Me_3Si is more strongly electron attracting than the trimethylgermyl group.

The ultraviolet (u.v.) absorption spectra and dipole moments of Me_3CPh, Me_3SiPh, Me_3GePh and Me_3SnPh and the corresponding benzyl derivatives Me_3MCH_2Ph have been examined (Nagy et al., 1967) and, like the e.s.r. spectra, these lead to the view that $d_\pi-p_\pi$ bonding is most pronounced between silicon and phenyl groups. The u.v. spectra show a hypsochromic shift from silicon to tin. Phenylchlorogermanes all show five bands in the ultraviolet that scarcely change in position or extinction coefficient in the series Ph_3GeCl, Ph_2GeCl_2, $PhGeCl_3$. The assignments of these transitions have been discussed by Marrot et al. (1965) in terms of inductive and conjugative factors.

Two additional techniques that have been applied to the question of π-bonding pose further problems. The work of Eaton and McClellan (1967) suggests that, even if π-bonding occurs between a group IV metal and a planar aromatic ring, the effect is not transmitted through the metal atom. Nickel(II) aminotroponimineates (1) exist in solution as an equilibrium mixture of the planar (diamagnetic) and tetrahedral (paramagnetic) forms.

In the absence of Ph_3M substituents about 10% of the unpaired electron density is delocalized over the π-system of the ligand, and this produces large contact shifts in the p.m.r. spectrum. If further delocalization over the three phenyl groups via a group IV metal substituted as in (1) occurred then further contact shifts should be observed. The experimental observations suggest that little or no conjugation effect exists for silicon, germanium or tin, but the energies of the phenyl, metal and ligand orbitals may be very different.

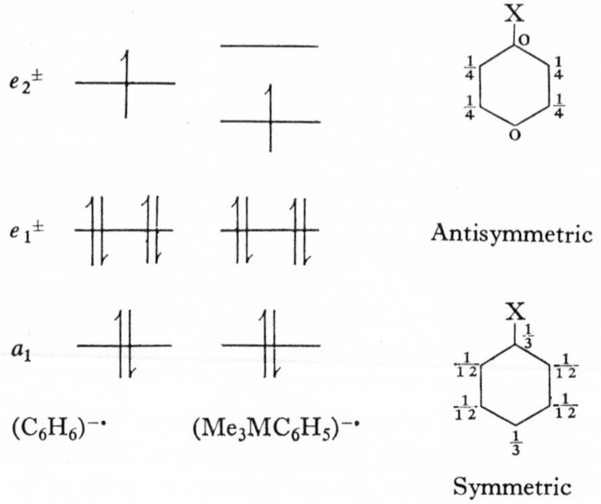

Ion	Proton		
	ortho	*meta*	*para*
Me$_3$CPh$^{-\cdot}$	4·66	4·66	1·74
Me$_3$SiPh$^{-\cdot}$	2·66	1·06	8·13
Me$_3$GePh$^{-\cdot}$	2·33	1·46	7·61

Fig. 1. Molecular orbitals of aromatic anions and proton hyperfine splitting constants.

Although the planarity of trisilylamine is most readily explained by silicon—nitrogen π-bonding, the force constants of the silicon—nitrogen bonds are more consistent with single-bond values. Nitrogen–proton coupling constants are useful parameters for examining the hybridization of nitrogen provided the dominant ^{15}N–H coupling is by the Fermi contact term. The magnitude of $J(^{15}$N–H) is then directly proportional to the s-character of the nitrogen atom directed towards the hydrogen atom to which it is bonded. The amount of s-character should increase with d_π–p_π bonding between nitrogen and a group IV metal, since this interaction involves either a pure p-orbital of nitrogen or one with a large p character. This technique has been applied to ^{15}N-enriched N-trimethylsilylaniline, $Me_3SiNHPh$ (and the germanium and tin analogues). In all three compounds, $J(^{15}N–H) = 76$ c/s, which is close to the value for sp^3 hybridized nitrogen in ammonium salts ([R$_3$NH]X) and very different from that in the

(1)

sp^2 hybridized pyridinium ion (90–100 c/s). Hence either the nuclear magnetic resonance (n.m.r.) hypothesis is incorrect, or there is no π-bonding (Randall and Zuckerman, 1966; Randall *et al.*, 1966).

III. THERMODYNAMIC DATA AND BOND LENGTHS

Heats of formation and bond energy data are available for only few germanium compounds, and Cottrell (1954) is critical of some of the bond-dissociation energies quoted. Even a recent determination of the heat of combustion of tetraphenylgermane leads to an unacceptably low value (32 kcal. mole^{-1}) for the mean germanium–carbon bond strength (Birr, 1962).

The heat of atomization of germanium $\Delta H_f^0(Ge, g)$ is reliably established as 90 ± 2 kcal. mole^{-1}. There is now good agreement on the heat of formation of germanium dioxide, and the most recent measurements by fluorine bomb calorimetry have established the thermochemical relationship between its three forms, hexagonal, tetragonal and glass (Jolly and Latimer, 1952b; Gross *et al.*, 1966)—

$$GeO_2 + 2F_2 \rightarrow GeF_4 + O_2$$

Comparison of the heats of formation of germanium- and silicon-tetra-halides (obtained by measuring their heats of solution in aqueous caustic soda) show that the stabilization of the metal—X bond due to such factors as ionic character and possible π-bonding is greater for silicon than germanium (Evans and Richards, 1952).

The heats of formation of germane, digermane and trigermane (Table II) have been measured by explosion in mixtures with stibine (Gunn and Green,

1961, 1964); average germanium—hydrogen and germanium—germanium bond strengths derived from these values are given in Table III.

TABLE II

Standard heats of formation

Compound	ΔH_f^0, kcal. mole^{-1}	Reference
GeO$_2$ (hexagonal)	$-132 \cdot 58 \pm 0 \cdot 19$†	
GeO$_2$ (tetragonal)	$-138 \cdot 66 \pm 0 \cdot 31$†	Gross et al. (1966)
GeF$_4$ (g)	$-284 \cdot 37 \pm 0 \cdot 15$†	
GeCl$_4$ (g)	-120†	
GeBr$_4$ (g)	$-83 \cdot 3$†	Cottrell (1954)
GeI$_4$ (g)	$-42 \cdot 0$†	
GeTe (s)	$-6 \pm 2 \cdot 3$‡	Colin & Drowart (1964)
GeH$_4$ (g)	$21 \cdot 7$†, $20 \cdot 8$‡	Gunn & Green (1961, 1964)
Ge$_2$H$_6$ (g)	$38 \cdot 8$†, $39 \cdot 1$‡	
Ge$_3$H$_8$ (g)	$54 \cdot 2$†, $48 \cdot 4$‡	Saalfeld & Svec (1963)
Et$_4$Ge (g)	$-34 \cdot 6 \pm 1 \cdot 2$†	
Pr$^n_4$Ge (g)	$-54 \cdot 9 \pm 1 \cdot 2$†	Pope & Skinner (1964)
Et$_6$Ge$_2$ (g)	$-78 \cdot 0 \pm 2 \cdot 5$†	Rabinovich et al. (1963)

† Calorimetric.   ‡ Electron impact.

Bond-energy data have also been derived from appearance-potential measurements on ions in the mass spectra of mono-, di- and tri-germane (Saalfeld and Svec, 1963), and the values are in good agreement for mono-

TABLE III

Average bond energies in germanium compounds

Bond	Energy, kcal. mole^{-1}	Compounds
Ge—H	68–69	GeH$_4$
Ge—Ge	38–39	Ge$_2$H$_6$, Ge$_3$H$_8$
	62 ± 5	Et$_6$Ge$_2$
Ge—C	57–59	Et$_4$Ge, Pr$^n_4$Ge
Ge—F	111	GeF$_4$
Ge—Cl	81	GeCl$_4$
Ge—Br	66	GeBr$_4$
Ge—I	51	GeI$_4$
Ge—Se	115	GeSe
Ge—Te	93	GeTe

and di-germane. Not all of the appearance-potential measurements reported are self-consistent (Table IV), which implies that some of the decomposition processes occur with excess kinetic energy, resulting in "tailing" of the ionization efficiency curve.

TABLE IV

Appearance potentials

Dissociation	Potential, eV
$GeH_4^{+\bullet} \rightarrow GeH_3^+ + H^\bullet + e$	$10\cdot8 \pm 0\cdot3$
$GeH_4^{+\bullet} \rightarrow GeH_2^{+\bullet} + H_2 + e$	$11\cdot8 \pm 0\cdot2$
$GeH_4^{+\bullet} \rightarrow GeH^+ + 3H^\bullet + e$	$16\cdot8 \pm 0\cdot3$
$GeH_4^{+\bullet} \rightarrow Ge^{+\bullet} + 2H_2 + e$	$10\cdot7 \pm 0\cdot2$
$GeH_4^{+\bullet} \rightarrow Ge^{+\bullet} + H_2 + 2H + e$	$14\cdot1 \pm 0\cdot5$
$GeH_4^{+\bullet} \rightarrow Ge^{+\bullet} + 4H^\bullet + e$	$18\cdot3 \pm 0\cdot3$

Heats of formation derived from appearance potentials are therefore likely to be low, and hence bond energies may represent upper limits. In the mass spectrum of monogermane, the reaction of lowest energy resulting in the formation of the $Ge^{+\bullet}$ ion is—

$$GeH_4 + e \rightarrow Ge^{+\bullet} + 2H_2 + 2e$$

The appearance potential, $A(Ge^{+\bullet})$ is given by

$$A(Ge^{+\bullet}) = \Delta H_f^0 (Ge^{+\bullet}) + 2\Delta H_f^0 (H_2) - \Delta H_f^0 (GeH_4) + E$$

In this expression, E represents the excess kinetic energy term, and if this is zero or small the standard heat of formation of germane may be obtained using available thermochemical data $[\Delta H_f^0 (Ge^{+\bullet}) = 267\cdot32$ kcal. mole$^{-1}]$.

Appearance-potential measurements on tetramethylgermane are less easily interpreted, since hydride ions are produced which may be involved in the threshold energy for a given process. Furthermore, the heat of formation of tetramethylgermane is unknown, but use of an estimated value $[\Delta H_f^0 (GeMe_4, g) = -35$ kcal. mole$^{-1}]$ leads to an average germanium—carbon bond strength of 64 kcal. mole^{-1} (Hobrock and Kiser, 1962).

Up to 1964 there was no reliable value for the germanium—carbon bond strength. Early work has been summarized by Bills and Cotton (1964), who measured the heat of combustion of tetraethylgermane in a rotating bomb calorimeter with 10% aqueous hydrofluoric acid as the bomb liquid $(\Delta E_c^0 = -1514\cdot1$ kcal. mole$^{-1})$. Using a value for the heat of vaporization of tetraethylgermane of $10\cdot1 \pm 0\cdot3$ kcal. mole^{-1}, this leads to $\Delta H_f^0(Et_4Ge, g)$

$= -39 \cdot 9 \pm 2 \cdot 0$ kcal. mole^{-1}, and hence to a mean thermochemical germanium—carbon bond energy of $58 \cdot 4 \pm 2$ kcal. mole^{-1}. Pope and Skinner (1964) have determined the standard heats of formation of both tetraethyl- and tetra-n-propyl-germane calorimetrically. Their values are listed in Table II, and average metal—carbon and metal—hydrogen bond strengths in Group IV metal-ethyls and -hydrides in Table V.

TABLE V

Average metal—carbon and metal—hydrogen bond strengths

MEt$_4$	Bond strength, kcal. mole^{-1}	MH$_4$	Bond strength, kcal. mole^{-1}
C—Et	82·1	C—H	99·3
Si—Et	~63	Si—H	77·4
Ge—Et	56·7	Ge—H	69·4
Sn—Et	46·2	Sn—H	60·4
Pb—Et	30·7	—	—

Two points of interest are apparent from Table V: one is the steady decrease in metal—carbon and metal—hydrogen bond strengths in going down the group, and the other is that metal—hydrogen bonds in MH$_4$ compounds are stronger than metal—carbon bonds in the tetraethyl derivatives. Rabinovich and co-workers (1963) have reported calorimetric measurements on tetraethylgermane and hexaethyldigermane. The latter ($\Delta E_c^0 = -2321 \cdot 0 \pm 2$ kcal. mole^{-1}) leads to a germanium—germanium bond strength of 62 kcal. mole^{-1}.

Appearance potentials have been used to compare bond strengths in several metal—metal bonded compounds (Chambers and Glockling, 1968). Although the accuracy of such measurements is not high, the following trends in relative bond energies have emerged—

$$Ph_3Sn—Ph > Ph_3Sn—GeMe_3 > Ph_3Sn—SnMe_3 > Ph_3Sn—SnPh_3$$

Bond lengths

Selected bond-length measurements on germanium compounds, derived from microwave spectra and X-ray crystallographic studies, are listed in Table VI. References should be sought in the appropriate part of the text.

A detailed analysis of the i.r. spectra of the Group IV hydrides, including monogermane, has been made using isotopes of the central atom and selective deuteration. In this way, bond distances have been calculated with an accuracy of $0 \cdot 001$ Å (Wilkinson and Wilson, 1966).

A microwave investigation of methylgermane, obtained from 28 isotopic forms by partial deuteration, reveals that although the germanium—carbon bond has an r_s length of 1·945 Å, the internal barrier to rotation of the

TABLE VI

Bond lengths

Bond	Length, Å	Compounds
Ge—H	1·529	H_3GeCN, H_3GeSiH_3, Cl_3GeH, $MeGeH_3$
Ge—F	1·68	GeF_4, $ClGeF_3$
	1·72	GeF_6^{2-}
	1·73	H_3GeF
	1·79	GeF_2 (shortest)
Ge—Cl	2·067	F_3GeCl
	2·09	$GeCl_4$
	2·11	$HGeCl_3$
	2·15	H_3GeCl
	2·26	$Cl_2Ge[Fe(CO)_3\pi-C_5H_5]_2$
Ge—Br	2·297	H_3GeBr, $GeBr_4$
Ge—I	2·55	H_3GeI
Ge—O	1·651	GeO
	1·90	$Na_4Ge_9O_{20}$ (GeO_6 octahedra)
Ge—C	1·919	H_3GeCN
	1·945	H_3GeCH_3, $Ph_3COGe—Ph_3$
	1·96	$I_2Ge \diagup \diagdown GeI_2$
	1·98	Me_4Ge, $Me_3—Ge—CN$
	2·01	$PhCO—GePh_3$
Ge—Ge	2·41	Ge_2H_6, Ge_3H_8
	2·45	Metal
Ge—Si	2·357	H_3GeSiH_3
Ge—Mn	2·54	$Ph_3GeMn(CO)_5$
Ge—Fe	2·36	$Cl_2Ge[Fe(CO)_3\pi-C_5H_5]_2$

methyl group remains fairly high at 1239 ± 25 cal. mole^{-1}. Quadruple coupling of the ^{73}Ge nucleus is very small, showing little asymmetry in the electrons at that nucleus (Laurie, 1959).

IV. VIBRATIONAL SPECTRA

Some of the simpler inorganic and organic compounds of germanium have been thoroughly examined, and assignments have been made for all or almost all of the fundamental vibrations. These are listed in Table VII; where more than one study has been reported, reference is made only to the more recent work. An example illustrating the frequencies of germanium–

TABLE VII
Infrared and Raman Spectra

Compound	Reference
GeH_4, GeD_4	Straley *et al.* (1942); Levin (1965); McKean & Chalmers (1967)
Ge_2H_6, Ge_2D_6	Crawford *et al.* (1962)
H_3GeD, D_3GeH	Lindeman & Wilson (1954)
H_3GeX, D_3GeX (X = F, Cl, Br, I)	Freeman *et al.* (1963); Griffiths (1967)
H_3GeN_3 $(H_3GeN)_2C$	Cradock & Ebsworth (1968)
H_3GeNCS, D_3GeNCS	Davidson *et al.* (1967)
H_3GeOMe, H_3GeOCD_3 H_3GeSMe $(H_3Ge)_2O$	Cradock (1968)
$(H_3Ge)_3P$	Cradock *et al.* (1967a)
H_3GeCN, D_3GeCN	Goldfarb (1962)
$MeGeH_3$, $MeGeD_3$, CD_3GeH_3, $MeGeD_3$	Griffiths (1963a)
$EtGeH_3$, $EtGeD_3$	Mackay & Watt (1967)
H_2GeX_2 (X = F, Cl, Br, I)	Ebsworth & Robiette (1964); Cradock & Ebsworth (1967)
Me_2GeH_2 Me_3GeH	Van der Vondel & Van der Kelen (1965b)
$HGeCl_3$, $DGeCl_3$	Lindeman & Wilson (1957)
GeF_4	Woltz & Nielsen (1952)
GeF_6^{2-}	Begun & Rutenberg (1967)
$GeCl_4$	Lindeman & Wilson (1957)
$GeCl_6^{2-}$, $GeCl_5^-$	Beattie *et al.* (1967)
$MeGeCl_3$	Aronson & Durig (1964)
Me_2GeCl_2	Griffiths (1964)
Me_3GeCl	Van der Vondel & Van der Kelen (1965a)
Me_4Ge	Lippincott & Tobin (1953)
$MeGe_2H_5$, $EtGe_2H_5$	Mackay *et al.* (1968b)
Me_6Ge_2	Brown *et al.* (1960a)
$Ge(C:CH)_4$	Sacher *et al.* (1967)
$Me_3GeC:CH$, $Me_3GeC:CD$ $Me_3GeC:CCl$	Steingross & Zeil (1966)
$Ge(NCO)_4$	Miller & Carlson (1961)
Me_3GeN_3	Thayer & West (1964)
$C_6H_5GeCl_3$, $C_6D_5GeCl_3$	Durig *et al.* (1966); Durig & Sink (1968)

hydrogen and germanium–halogen vibrations is given in Table VIII. One noteworthy feature is the change in germanium–hydrogen stretching frequency between fluorides, chlorides and bromides.

The i.r. spectra of ethylgermane and ethyltrideuterogermane, in conjunction with their Raman spectra, allow the assignment of 24 fundamental vibrations. In the solid state, ethylgermane has one strong broad adsorption due to ν(Ge–H) at 2062 cm^{-1}, and one broad band is observed in the

TABLE VIII

Infrared spectra of dihalogermanes, cm^{-1}

		H_2GeF_2	H_2GeCl_2	H_2GeBr_2
ν_6	GeH$_2$ (antisym. stretch)	2174·4	2150·3	2138·3
ν_1	GeH$_2$ (sym. stretch)	2154·5	2134·6	2121·4
ν_9	GeX$_2$ (antisym. stretch)	720	435	322
ν_3	GeX$_2$ (sym. stretch)	720	410	290
ν_2	GeH$_2$ (scissors)	860·0	854·5	847·3
ν_8	GeH$_2$ (wag)	813·5	779·4	754·0
ν_5	GeH$_2$ (twist)		Forbidden in i.r.	
ν_7	GeH$_2$ (rock)	596	524	489

Raman spectrum at 2070 cm^{-1}. The i.r. spectrum of the gas shows a strong band with three maxima, and the three germanium–hydrogen stretching modes are included in this one contour centred on 2085 cm^{-1}.

So many papers deal at least superficially with the i.r. spectra of polyatomic germanium compounds that it is impractical to quote every example; many papers simply list frequencies and intensities. Two key references (Cross and Glockling, 1965a; Ulbricht and Chvalovsky, 1968) provide examples covering much of the field.

A. Hydrides and deuterides

Many complex organogermanium hydrides and deuterides have been examined (Mathis *et al.*, 1962, 1964a; Massol and Satge, 1966). In triphenylgermane and its deuteride, stretching and bending frequencies are: ν(Ge–H), 2037; δ(Ge–H), 709; ν(Ge–D), 1473; and δ(Ge–D), 526 cm^{-1}. Germanium–hydrogen stretching frequencies cover the range 2175 [in H_2GeF_2] to 1953 cm^{-1} [in $(Ph_3Ge)_3GeH$]. For alkyl and related groups bonded to germanium, the germanium–hydrogen stretching frequency has been related to the sum of the Taft σ^* coefficients of the attached groups—

$$\nu(\text{Ge–H}) = 2008 + 16 \cdot 5 \Sigma \sigma^*$$

The integrated germanium–hydrogen absorption intensity is enhanced by electron attracting $(-I)$ substituents on germanium, whereas electron-repelling substituents decrease the intensity.

B. Halides

Germanium–halogen stretching frequencies in complex molecules cover a wide range. For fluorides, frequencies between 485 and 608 cm^{-1} have been quoted (Jezowska *et al.*, 1967), whereas several chlorides (R_3GeCl) absorb between 362 and 379 cm^{-1}. Bands ascribable to germanium–bromine stretching frequencies in R_3GeBr, R_2GeBr_2 and $RGeBr_3$ compounds occur in the range 251–327 cm^{-1} and germanium tetrabromide has a strong band at 330 cm^{-1}. Germanium tetraiodide absorbs strongly at 263 cm^{-1} and triphenyliodogermane and methyltri-iodogermane absorb in this region (283 and 256 cm^{-1}). The Raman spectra of $EtGeCl_3$ and Et_2GeCl_2 have been described (Lippincott *et al.*, 1953).

C. Oxides

Germanium dioxide absorbs most strongly at 880 cm^{-1}, and many organogermanes of the types $(R_3Ge)_2O$ and $(R_2GeO)_n$ are similar, absorbing in the region 875–926 cm^{-1}. Alkoxygermanes, $Ge(OR)_4$, show two bands at 1040 and 680 cm^{-1}. Germanium–oxygen stretching vibrations are always broad and in cyclic oxides, such as $(Me_2GeO)_n$, changes are observed depending on whether the trimer, tetramer or polymeric form is present (Brown *et al.*, 1960b). Typical values are: $(Et_3Ge)_2O$, 855; $(Ph_3Ge)_2O$, 858; and $[Et(H)GeO]_4$, 870 cm^{-1}. In the phosphate ester, $Et_3Ge—OP(O)Ph_2$, a much higher frequency (954 cm^{-1}) is observed.

D. Sulphides

In trimeric sulphides, $(R_2GeS)_3$, symmetric and asymmetric germanium-sulphur stretching frequencies have been reported ($\nu_{asym.} \sim 400–420$; $\nu_{sym.}$, 300–370 cm^{-1}) (Schumann, 1967).

E. Phosphides

In Et_3GePPh_2, the germanium–phosphorus stretching frequency has been ascribed to a band at 474 cm^{-1}. Lower frequencies have been reported for compounds such as $(Ph_3Ge)_3P$ ($\nu_{asym.}$, 392; $\nu_{sym.}$, 368 cm^{-1}).

F. Alkylgermanes

Bands due to germanium—carbon stretching vibrations are observed near 600 cm^{-1}, and in many compounds both the asymmetric and symmetric vibrations are observable (in Ph_2GeEt_2 these occur at 579 and 553 cm^{-1}, respectively). The i.r. spectra of n-propyl- and n-butyl-germanes provide evidence for *trans* and *gauche* forms with ν(Ge–C) *trans* near 645 and ν(Ge–C) *gauche* near 550 cm^{-1} (Carrick and Glockling, 1966). In tetra-alkylgermanes, the germanium–carbon stretching frequency moves to lower values in the series $Me_4Ge > Et_4Ge > Pr^i_4Ge$, but for Pr^n_4Ge, Bu^i_4Ge and Bu^n_4Ge, a shift to higher frequency is observed. In hexa-alkyldigermanes ν(Ge–C) is lower than in the corresponding monogermanes (Glockling and Light, 1967).

Identification of individual alkyl groups in organogermanes is often possible. For example, the methyl rocking mode in methylgermanes occurs in the range 850–787 cm^{-1} and is usually one of the strongest bands present. Ethyl-, propyl- and butyl-germanes also show diagnostically significant bands connected with the vibrations of the hydrocarbon groups. In the series $Pr^n_xGePr^i_{4-x}$ ($x = 0$–4) progressive changes in the carbon–hydrogen stretching frequencies are of diagnostic value. For n-butylgermanes two weak bands near 880 and 870 cm^{-1} are characteristic, and benzylgermanes have a group of bands near 815, 805 and 775 cm^{-1} that probably involve a CH_2 rocking mode. In 4- and 5-membered germanacyclanes, a band at 1120 cm^{-1} is evidently characteristic of these systems.

Carbon—carbon double and triple bonds when attached to germanium (GeC:C and GeC:C) show stretching frequencies that are generally about 60 cm^{-1} lower than in similar carbon compounds (Egorov *et al.*, 1962). In alkynylgermanes, $R_3GeC:CR'$ the Taft σ^* coefficient of R' shows a linear relationship to the intensity of the ethynyl stretching vibration (Mathis *et al.*, 1964b).

G. Arylgermanes

Phenylgermanes produce i.r. bands characteristic of monosubstituted benzenes. Harrah *et al.* (1962) observed an "X-sensitive" band at 1089 cm^{-1} that moves to lower frequency in phenyl-tin and -lead compounds (Whiffen, 1956). For a wide variety of phenylgermanes, this band, which is strong and sharp, varies only slightly [1099 in $(Ph_3Ge)_2O$ to 1081 cm^{-1} in $PhGeBr_3$], but in some metal—metal bonded complexes, such as $(Ph_3Ge)_2Pt(PEt_3)_2$, it moves to lower frequency (1073 cm^{-1}). An even more constant band is observed at 1428 cm^{-1}, the observed variations being no greater than 5 cm^{-1}. A low-frequency band at 450 cm^{-1}, which is reported as splitting in the more unsymmetrical phenyl derivatives of the Group IV metals, varies over the range 482–448 cm^{-1}.

Phenylgermanium compounds show two bands in the region 350–200 cm^{-1}, of which the stronger, near 325 cm^{-1}, is highly characteristic of the triphenylgermyl group and falls at lower frequency (314–309 cm^{-1}) in diphenylgermyl compounds. It is probable that this vibration includes the asymmetric germanium–phenyl stretching frequency. The weaker band, which may include the symmetric germanium–phenyl stretch is more variable in position (303–268 cm^{-1}) (Mackay *et al.*, 1968a).

H. Germanium—metal Bonds

Little information is available on metal-bonded compounds. The Ge–Ge stretching frequency in (Ph$_3$Ge)$_3$GeH and (Ph$_3$Ge)$_3$GeMe is believed to occur near 228 cm^{-1}. A band at 351 cm^{-1} in the spectrum of (PhCH$_2$)$_3$-GeSiMe$_3$ has tentatively been assigned to ν(Ge–Si). In digermane and hexadeuterodigermane, ν(Ge–Ge) is observed at 269·8 and 261·5 cm^{-1}, respectively, in their Raman spectra.

V. NUCLEAR MAGNETIC RESONANCE SPECTRA

Many of the papers concerned with organogermanes or germanium hydrides that have been published in the past few years contain p.m.r. data. Aspects relevant to the establishment of the structure of a given compound

TABLE IX

Proton chemical shifts† and coupling constants

Compound	GeH$_3$	GeH$_2$	GeH	CH$_3$	J(GeH$_3$–GeH$_2$)	J(GeH$_2$–CH$_3$)
Me$_3$GeH	—	—	6·08	9·79	—	—
Me$_2$GeH$_2$	—	6·27	—	9·71	—	—
MeGeH$_3$	6·51	—	—	9·65	—	—
Me(H)$_2$GeGeH$_3$	6·79	6·42	—	9·79	3·9	4·3
Cl(H)$_2$GeGeH$_3$	6·37	4·61	—	—	4·1	—
Br(H)$_2$GeGeH$_3$	6·21	5·31	—	—	4·1	—
I(H)$_2$GeGeH$_3$	5·96	6·57	—	—	4·3	—

† Relative to internal TMS.

are described in later Chapters. Chemical shifts and coupling constants have been tabulated for the period up to 1964 (Maddox *et al.*, 1965). Germanium has one magnetic isotope and only in liquid tetramethylgermane has splitting of the methyl resonance by the ^{73}Ge nucleus ($I = \frac{9}{2}$, 7·6% abundant) been observed. The value of the spin–spin coupling constant [$J(^{73}$Ge–^{1}H) = 2·94 c/s] has been obtained from the outer three lines of the theoretical 10-line spectrum (Tzalmona, 1963).

Chemical shifts and coupling constants for some representative germanium hydrides are given in Table IX. For the digermanes, Ge$_2$H$_5$X, the spectrum is simple first-order, consisting of a triplet (GeH$_3$) and a quartet

(GeH$_2$) (Mackay *et al.*, 1966, 1968b). In methylgermanes, the hydride resonance shifts to low field with increasing substitution of methyl groups in monogermane, whereas the methyl proton resonance moves to high field. In this series, spin–spin coupling constants are of the order of 4 c/s (Schmidbaur, 1964). Successive halogenation of tetramethylgermane results in a progressive down-field shift in the proton resonance, and the magnitude increases in the order Cl < Br < I. Linear correlations have been reported between chemical shifts and the sum of Taft σ^* coefficients of attached groups, but these may be largely anisotropy effects (Egorochkin *et al.*,

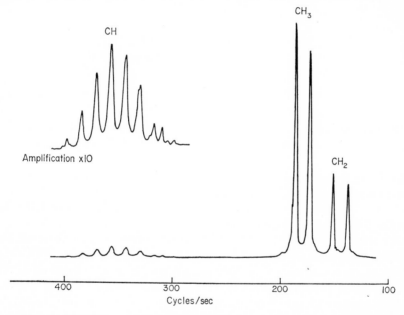

Fig. 2. 100 Mc/s p.m.r. spectrum of Bu$^i_4$Ge (downfield from TMS).

1964b; Schmidbaur and Ruidisch, 1964). In bistrimethylgermylmethane, (Me$_3$Ge)$_2$CH$_2$, the methylene protons are at unusually high field (10·13τ), whereas in the substituted ethane the methylene resonance lies downfield from TMS (9·28τ) (Egorochkin *et al.*, 1964a). The p.m.r. spectrum of tetraisobutylgermane, (Me$_2$CHCH$_2$)$_4$Ge, in carbon tetrachloride at 33°C is shown in Fig. 2. The methyl protons form a doublet (8·20τ; J(CH$_3$–CH) = 13·0 c/s) and the methylene protons likewise form a doublet [8·55τ; J(CH$_2$–CH) = 14·3 c/s]. The closeness of these two coupling constants results in the methyne proton appearing (at 100 Mc/s) as an 8-line resonance (Glockling and Light, 1967). The p.m.r. spectrum of melts or concentrated solutions of dimethylgermanium oxide consists of a single broad peak that cannot be ascribed to viscosity effects, but rather to rapidly exchanging species at equilibrium. The average lifetime of an exchanging species in the melt at 100°C has been estimated as about 0·3 sec—

$$4(Me_2GeO)_3 \rightleftharpoons 3(Me_2GeO)_4$$

The proton shift of the cyclic trimer (−0·495 p.p.m. from TMS) differs from the value of −0·472 for the cyclic tetramer (Moedritzer, 1966b).

Hexavinyldigermane shows a partially collapsed ABC pattern for the vinyl protons with coupling constants close to those in tetravinyl- and triphenylvinyl-germane—

$$\begin{array}{c} H_A \\ H_B \end{array} C = C \begin{array}{c} H_C \\ Ge \end{array}$$

$J(AB) = 2\cdot93$ c/s	$\delta A\dagger = -335\cdot56$
$J(AC) = 20\cdot07$ c/s	$\delta B = -357\cdot46$
$J(BC) = 13\cdot32$ c/s	$\delta C = -371\cdot61$
$\Sigma J = 36\cdot32$ c/s	

† p.p.m. relative to TMS.

Various authors have reported a linear correlation between the sum of $J(cis)$, $J(trans)$ and $J(gem)$ and the electronegativity of the substituent group. For vinylsilanes, the sum is 38·8 c/s; hence if this criterion is applied (and its validity is suspect) germanium is more electronegative than silicon in these compounds (Cawley and Danyluk, 1963).

In trimethylmetal sulphides, Me_3MSMe and $(Me_3M)_2S$ (where $M = C$, Si, Ge, Sn or Pb) the chemical shift of the methyl-metal protons follows the order $Si > Sn > Ge > Pb > C$, which is the same as that observed for Me_4M compounds (Allred and Rochow, 1958; Abel and Brady, 1968). The effect of electronegative S or SMe groups results in a deshielding of the methyl-metal protons compared to the tetramethyls. However the de-shielding effect in $(Me_3M)_2S$ compounds is greater than in $(Me_3M)_2O$, which is not what is expected on electronegativity grounds alone. This apparent anomaly is probably due to induced currents in the localized electrons of the metal—sulphur bonds; anisotropic deshielding by the larger sulphur atom being greater than the inductive deshielding of the oxygen compounds. Methylgermanium sulphides, $MeGe(SMe)_3$ and $Me_2Ge(SMe)_2$, have been compared (Van den Berghe et al., 1967) and the downfield chemical shift of the Me-Ge protons is found to increase linearly with the number of thiomethyl groups. The sulphur-methyl proton shifts move upfield, again linearly, with increasing number of sulphur-methyl groups.

Fluorine-19 magnetic resonance spectra of perfluorovinylgermanes have been reported (Coyle et al., 1961; Seyferth et al., 1962b). In these compounds each fluorine atom is magnetically distinguishable and the spectra of $CF_2:CFGeR_3$ compounds consist of three quartets with all lines of equal intensity. Coupling constants are in the order $J(trans) > J(gem) > J(cis)$ and the values are sufficiently characteristic for structural assignments to be

TABLE X

^{19}F chemical shifts and coupling constants

Compound	Chemical shift,† p.p.m.			Coupling constants, c/s		
	δ_1	δ_2	δ_3	$J(trans)$	$J(cis)$	$J(gem)$
$(CF_2:CF)_4Ge$	80·1	112·7	196·5	118	32	71
$(CF_2:CF)_2GeMe_2$	86·6	118·6	195·5	118	32	72
$(CF_2:CF)GeEt_3$	90·2	121·9	194·4	115·5	31·9	78·8
$Ph_3GeCF:CFSiEt_3$	—	154·5	158·5	144	—	—

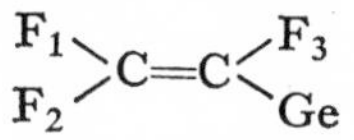

$$\begin{array}{c} F_1 \\ F_2 \end{array}\!\!\!>\!C\!=\!C\!<\!\!\!\begin{array}{c} F_3 \\ Ge \end{array}$$

† δ relative to internal CCl_3F, positive shifts to high field.

made as in the compound *trans*-$Ph_3GeCF:CFSiEt_3$ (Seyferth and Wada, 1962). Typical values of chemical shifts and coupling constants are given in Table X.

VI. MASS SPECTRA

A short review has been published on the mass spectra of organometallic compounds (Chambers *et al.*, 1968). Mass-spectroscopic examination of germanium compounds is complicated by the polyisotopic nature of the metal (Table XI), which produces a characteristic pattern for each germanium-containing ion.

TABLE XI

Mass and abundances of germanium isotopes

	Mass (^{12}C scale)	Percentage abundance
^{70}Ge	69·924277	20·56
^{72}Ge	71·921740	27·42
^{73}Ge	72·923360	7·79
^{74}Ge	73·921150	36·47
^{76}Ge	75·921360	7·76

For ions containing more than one germanium atom, the pattern changes in a way that can assist in the identification of species under low resolution. This is illustrated for ions derived from isopropylgermanium oxides containing 2, 3 and 4 germanium atoms (Fig. 3) (Carrick and Glockling, 1966, 1967).

One complication in the interpretation of spectra is that fragmentation processes that result in loss of a small neutral fragment produce overlapping patterns. In these cases, positive identification of each species may require high-resolution mass measurements on selected ions. Metastable peaks show the isotope pattern of the parent ion if the transition involves elimination of a neutral monoisotopic fragment. When a polyisotopic-containing

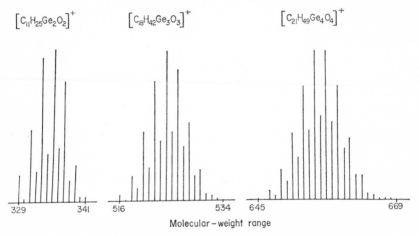

$$\left[C_{11}H_{25}Ge_2O_2\right]^+ \qquad \left[C_{18}H_{42}Ge_3O_3\right]^+ \qquad \left[C_{21}H_{49}Ge_4O_4\right]^+$$

329   341   516   534   645   669

Molecular—weight range

FIG. 3. Calculated isotope patterns for ions derived from $(Pr^i_2GeO)_4$.

fragment is eliminated as a neutral molecule or radical, the shape of the metastable peaks is profoundly altered as in the following illustrations (Fig. 4).

Few inorganic germanium compounds have been studied. In germanium tetrachloride, ion abundances are as follows: $GeCl_4^{+\cdot}$, 16·9; $GeCl_3^+$, 73·1; $GeCl_2^{+\cdot}$, 1·3; $GeCl^+$, 6·4; and $Ge^{+\cdot}$, 1·2%, and two doubly charged ions ($GeCl_3^{2+\cdot}$, $GeCl^{2+\cdot}$) are also present. Overlapping fragmentation patterns create the greatest difficulty with germanium hydrides owing to loss of H and H_2. Saalfeld and Svec (1963) overcame these difficulties by examining the monoisotopic hydrides $^{74}GeH_4$, $^{74}Ge_2H_6$ and $^{74}Ge_3H_8$ under low resolution and showed that the molecular ion for monogermane is of low abundance, although for digermane it is the fourth most abundant ion. For both di- and tri-germanes a high proportion of the ion current is carried by Ge_2- and Ge_3-containing ions, respectively. A more recent study of monogermane has shown the presence of two doubly charged ions ($GeH^{2+\cdot}$, Ge^{2+}) and a metastable peak for hydrogen elimination—

$$GeH_2^{+\cdot} \rightarrow Ge^{+\cdot} + H_2$$

The first mass-spectrometric examination of an organogermane was reported by Dibeler (1952), and quite a wide variety of compounds have since been examined (Glockling and Light, 1968; Chambers and Glockling, 1968; Duffield *et al.*, 1968). Most of the ion current is carried by germanium-containing species and the fragmentation behaviour of most compounds is

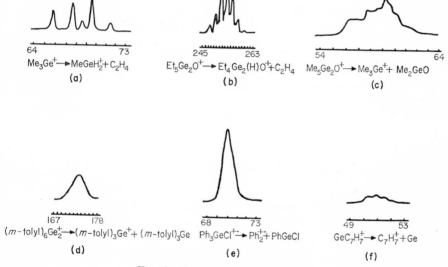

FIG. 4. Shapes of metastable peaks.

dominated by the low abundance of odd-electron ions. For example, molecular ions are usually of low abundance, which is consistent with the view that a *bonding* electron is removed on ionization. One notable exception is $(C_6F_5)_4Ge$, for which the parent ion is the second most abundant in the spectrum (Miller, 1967). In tetraphenylgermane, the molecular ion is some 20 times more abundant than in tetraphenyltin, although the Ph_3M^+ abundances are almost identical; this may be a reflection of the greater germanium—phenyl bond strength. Similarly, odd-electron fragment ions are generally of low abundance except in cases where there exists a particularly favourable process leading to their formation, e.g.—

$$R_3GeH^{+\cdot} \rightarrow R_2Ge^{+\cdot} + RH$$
$$Ph_4Ge^{+\cdot} \rightarrow Ph_2Ge^{+\cdot} + Ph_2$$

Parent ions are observed in almost all of the compounds examined. Some exceptions are triphenyliodogermane and cyclic oxides $(R_2GeO)_{3,4}$. Molecular ions largely decompose by cleavage of a germanium—X bond with elimination of a radical and formation of an even-electron product ion, and these processes are frequently metastable supported, e.g.—

$$Me_4Ge^{+\cdot} \to Me_3Ge^+ + Me^\cdot$$

Unsymmetrical germanes of the types A_3GeB and A_2GeB_2 always produce ions corresponding to cleavage of both germanium—A and germanium —B bonds, though the abundances of the product ions vary greatly—

$$Ph_3GeEt^{+\cdot} \begin{cases} \nearrow Ph_3Ge^+ + Et^\cdot \quad (64\%) \\ \searrow Ph_2EtGe^+ + Ph^\cdot \quad (0\cdot8\%) \end{cases}$$

Such variations may be the result of differences in bond strengths in the molecular ion, but the resultant ions will usually be able to decompose by different routes, so that conclusions concerning relative bond strengths are not necessarily valid.

Alkylgermanes having the grouping R_2CHCH_2Ge show that alkene elimination is a favoured process of low activation energy that only occurs from even-electron ions—

$$Et_3Ge^+ \to Et_2GeH^+ \to EtGeH_2^+ \to GeH_3^+$$

In many cases successive loss of alkene is metastable supported and the even-electron character is retained throughout. This type of reaction is probably a β-elimination process—

$$Et_2Ge^+ \underset{\overset{\displaystyle|}{H}}{\overset{\displaystyle CH_2}{\diagup}}\diagdown_{CH_2} \longrightarrow Et_2GeH^+ + C_2H_4$$

There is also evidence that ethylgermanium ions lose methylene in a unimolecular reaction which has been more fully studied for ethyltin compounds (Chambers *et al.*, 1967)—

$$Et_3Ge^+ \to Et_2GeCH_3^+ + CH_2$$

The proportion of hydrocarbon ions increases markedly with the length of the carbon chain in the order Me < Et < Pr < Bu.

Fragmentation of phenyl groups in phenylgermanes is a process of high activation energy which, at 70 eV, produces ions owing to successive elimination of acetylene and hydrogen. This is illustrated by the fragmentation diagram for triphenylgermane (Fig. 5). In this and the following figures, solid arrows denote the elimination of even-electron fragments and broken arrows the elimination of odd-electron fragments. Ion abundances are shown under the formulae, and metastable-supported transitions are indicated by insertion of the neutral fragment against the arrow.

Compounds containing a germanium—germanium bond or germanium bonded to another metal show additional fragmentation routes. In cases such as hexaethyldigermane, loss of an ethyl radical is followed by successive ethylene elimination, so that a considerable proportion of the ion current is carried by Ge_2 species (Fig. 6). The position is very different in digermanes,

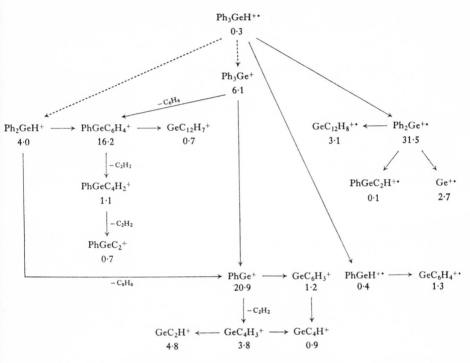

FIG. 5. Fragmentation of triphenylgermane.

such as Me_6Ge_2 and Ph_6Ge_2, that cannot undergo analogous alkene elimination reactions. In these compounds cleavage of the germanium—germanium bond in the ions $R_6Ge_2^{+\cdot}$ and $R_5Ge_2^+$ must be the processes of lowest activation energy (Fig. 7). Compounds containing germanium bonded to another metal also produce ions derived from the transfer of organic groups (phenyl and methyl) from one metal to the other. It is possible that these ions are all formed from the molecular ion by rearrangement followed by cleavage of the metal—metal bond. For example, compounds of the type $A_3MM'B_3$ show all six rearrangement ions—

$$A_3MM'B_3^{+\cdot} \rightarrow A_2MB^+ + AMB_2^+ + MB_3^+ + B_2M'A^+ + BM'A_2^+ + A_3M'^+$$

2

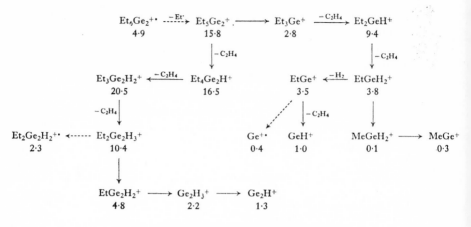

FIG. 6. Fragmentation of hexaethyldigermane.

By contrast, the hydrocarbon Ph_3CCMe_3 does not produce rearrangement ions, but this may be due to the large differences between the weak central carbon—carbon bond strength and all others in the molecule. These

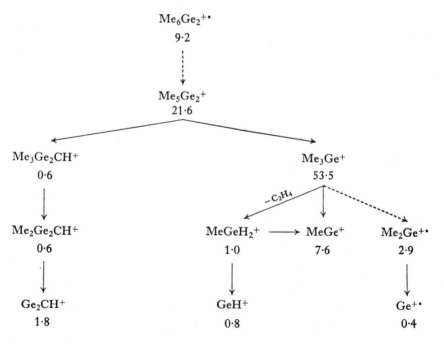

FIG. 7. Fragmentation of hexamethyldigermane.

metal—metal bonded compounds reveal metastable peaks for the elimination of neutral metal-containing molecules and radicals—

$$Ph_3Me_2GeSn^+ \rightarrow Ph_2GeMe^+ + PhSnMe^{\cdot\cdot}$$
$$(PhCH_2)_2GeSiMe_3^+ \rightarrow PhCH_2Ge^+ + PhCH_2SiMe_3$$

Appearance-potential measurements demonstrate that the germanium—tin bond in $Ph_3SnGeMe_3$ is some 12 kcal. mole^{-1} stronger than the phenyl—tin bond in tetraphenylstannane. Several higher methylpolygermanes have been examined under electron impact (F. Glockling and J. R. C. Light, unpublished work), and these produce simple spectra in that loss of a methyl radical is followed by successive elimination of Me_2Ge units. A complicating feature is the occurrence of rearrangement processes in which methyl groups are transferred from one germanium atom to another.

2 | INORGANIC CHEMISTRY

I. THE METAL, BINARY METALLICS

Germanium was first isolated from argyrodite, $4AgSGeS_2$, as the sulphide GeS_2, but other less rare minerals now provide the main source. Of these germanite, from Tsumeb in S.W. Africa contains 10% germanium, whilst reinerite (Steygers, 1960), found in the Congo at Kipushi, contains $6\cdot4-7\cdot8\%$ germanium as the disulphide. Germanium is also obtained from various zinc ores containing $0\cdot01-0\cdot1\%$ germanium, and is isolated in the smelting process. Some coal ashes and flue dusts yield up to 1% of germanium (Morgan and Davies, 1937).

Germanium-containing ores or extracts are treated with concentrated hydrochloric acid and distilled giving the volatile germanium tetrachloride, which is only sparingly soluble in strong hydrochloric acid. Re-distillation from aqueous hydrochloric acid saturated with chlorine (to remove arsenic as H_3AsO_4) followed by hydrolysis with aqueous ammonia gives relatively pure germanium dioxide. Reduction of the dioxide is most commonly carried out by carbon or a mixture of carbon and potassium cyanide at 1200°C (Jaffee *et al.*, 1946); these reduction processes are complicated by the volatility of germanium(II) oxide above 750°C. Final purification of the metal is effected by zone melting (Billig, 1955). Electrodeposition of germanium from fused-salt mixtures has been carried out, as also has the reduction of the dioxide by hydrogen at 900°C (Andrieux and Andrieux, 1955; Monnier and Tissot, 1964).

Germanium, like silicon, crystallizes in the diamond lattice. It is brittle with a bright lustre and is stable to water, 50% sodium hydroxide, concentrated hydrochloric and sulphuric acids. The metal is tarnished by aqueous caustic soda or concentrated nitric acid; it dissolves in 3% hydrogen peroxide with precipitation of the dioxide. Table XII lists the more important properties of the metal.

Germanium forms compounds or eutectics with a wide range of metals or metalloids. All of the alkali metals form compounds M-germanium. Sodium germanide is a hard, dense, pyrophoric material which is monoclinic, whereas KGe, RbGe and CsGe are cubic. Rubidium germanide forms bronze crystals and caesium germanide is black (Schaefer and Klemm, 1961; Busmann, 1961). The magnesium–germanium system shows two eutectics and the compound Mg_2Ge. Calcium and germanium react at a red heat forming a dark grey crystalline powder, CaGe. Aluminium and germanium form a eutectic, m.p. 423°C, containing 52% germanium. The addition of small amounts of germanium to aluminium alloys improves hardness and rolling properties. Germanium and silicon are evidently completely miscible in solid solution, whereas germanium and lead are only

TABLE XII

Properties of germanium

Atomic number	32
Mean atomic weight (^{12}C scale)	72·59
Density (25°C)	5·323
Melting point (°C)	937·4
Boiling point (°C)	2830
Heat of fusion (kcal./g atom)	8·1
Heat of vaporization (kcal./g atom)	79·9
Hardness (Mohr's scale)	6·0
Ionization potential, 1st (eV)	8·09
Ionization potential, 2nd (eV)	15·86
Ionization potential, 3rd (eV)	34·07
Ionization potential, 4th (eV)	45·5
$Ge^{2+} + 2e \rightarrow Ge$, $E°$ (volts)	0·0

completely miscible in the liquid state (Briggs and Benedict, 1930). GeP and GeSe binary compounds have been described, the latter having a distorted sodium chloride lattice (Zumbusch *et al.*, 1948; Dutta and Jeffrey, 1965). Silver produces a eutectic alloy containing 25 atom% germanium, melting point at 650°C, and gold also forms a gold-coloured, low-melting alloy (356°C) with germanium that is stronger but less ductile than gold. It has good soldering and casting characteristics. The crystal structure of Pd_2Ge and Pt_2Ge have been determined (Anderko and Schubert, 1953).

II. ANALYTICAL METHODS

Inorganic germanium compounds may be analysed for germanium by precipitation of the sulphide, GeS_2, from quite strongly acid solution. This method has been criticized because some reduction to the insoluble mono-sulphide occurs, and a more accurate method therefore consists in igniting the precipitated sulphide in an open crucible to the white dioxide. Germanium may also be precipitated from dilute sulphuric acid solution as the tannin complex, which can be ignited to GeO_2 (Holness, 1948). Volumetric methods have been described based on the formation of acid complexes, such as mannitol germanic acid (Bradley *et al.*, 1956) and pyrocatechol germanic acid (Wunderlich and Gohring, 1959), which may be titrated with a strong base. Several photometric methods are available based on complex formation, with for example, 2,3,7-trihydroxy-6-fluorone (Kazarinova and Vasilera, 1958).

In the analysis of organoderivatives of germanium, the difficulty is to destroy the organic groups without loss of germanium. Arylgermanes present no great problem: heating with fuming nitric and sulphuric acids, preferably with the addition of ammonium persulphate, followed by ignition at 700–1000°C, brings about quantitative conversion to germanium dioxide. Alkylgermanes, especially the more volatile methyl and ethyl derivatives, present difficulties owing to their high volatility and resistance to complete oxidative degradation. For example, prolonged boiling of dibenzyldimethylgermane with the above acid mixture gave the germanium content 10–20% low, presumably owing to loss of the volatile dimethyl-germanium oxide. The only satisfactory degradative method in these cases consists in fusion with sodium perchlorate in a steel bomb. In the determination of carbon and hydrogen, organogermanes commonly require a higher combustion temperature than organic compounds.

III. GERMANIUM(II) COMPOUNDS

Whereas with silicon the +2 oxidation state is of very low stability, it is well defined with germanium in various inorganic compounds. There is however no evidence for the existence of uncomplexed Ge^{2+} ions in solution, in contrast to tin and lead, and germanous compounds are quite strongly reducing in aqueous solution.

A. Germanium(II) oxide, GeO

This was first prepared by Dennis and Hulse (1930) by the reduction of germanium dioxide with hypophosphorous acid in hydrochloric acid solution. The immediate product, $Ge(OH)_2$, is precipitated as a yellow–red solid that darkens on heating until, at 650°C in a nitrogen atmosphere, it yields the anhydrous monoxide as jet-black needles. It sublimes above 700°C and may also be obtained by volatilization from an intimate mixture of germanium and germanium dioxide at 800°C (Mueller et al., 1926), although the phase diagram of the GeO_2–Ge system does not show the existence of a GeO phase. The monoxide is also formed by evaporation of the metal at 700°C in a stream of carbon dioxide at 1 atm or in an air stream at reduced pressure (Gastinger, 1956).

Germanium(II) oxide is stable to moist air at room temperature, but is oxidized in air at 550°C. As the anhydrous solid it is apparently insoluble in and unaffected by hydrochloric acid, sulphuric acid, and aqueous caustic soda. Its heat of formation, $\Delta H^0_{f\,298} = 53{\cdot}7$ kcal. mole^{-1} has been derived from vapour-pressure measurements on the system (Jolly and Latimer, 1952b)—

$$\tfrac{1}{2}Ge(s) + \tfrac{1}{2}GeO_2(s) \rightleftharpoons GeO(g)$$

It reacts with hydrogen chloride, chlorine and bromine—

$$GeO + 3HCl \xrightarrow{175°C} HGeCl_3 + H_2O$$
$$2GeO + 2X_2 \longrightarrow GeO_2 + GeX_4$$

The stability of GeO contrasts with that of SiO, which exists only in the vapour phase at high temperature.

B. Germanium(II) hydroxide, Ge(OH)₂

This compound is formed by the reduction of germanium dioxide with hypophosphorous acid in hydrochloric acid solution when, by the addition of ammonia, $Ge(OH)_2$ is precipitated as a yellow solid, readily oxidized by air (Dennis and Hulse, 1930). It is also formed by the hydrolysis of germanium(II) chloride. The colour of $Ge(OH)_2$ can vary from white to dark brown depending on the detailed precipitation procedure. Its i.r. spectrum supports the dihydroxy structure rather than the hydrated oxide (Dupuis, 1960). It forms a true solution in perchloric acid, whereas in strongly basic media it gives deep red colloidal solutions (Gayer and Zajicek, 1964a).

C. Germanium(II) sulphide

This may be prepared by the action of hydrogen sulphide on an acid solution of germanium(II) chloride, but the product contains up to 20% of chloride (Foster, 1946; Schumb and Smyth, 1955). It has also been obtained by reduction of the disulphide (Dennis and Joseph, 1927)—

$$GeS_2 \xrightarrow{H_2 \text{ or } Ge} GeS$$

Germanium monosulphide forms black crystals having a distorted sodium chloride type structure. Precipitation from solution (e.g., H_2S on $GeCl_2$) gives a dark-red amorphous solid that is converted into the black crystalline form at 450°C in an inert atmosphere. It is of high thermal stability, melting at 530°C to a black liquid having a vapour pressure of 2 mm at 635°C (Davydov and Diev, 1957). It dissolves in hydrochloric acid, caustic soda or yellow ammonium sulphide, and is oxidized by air at 350°C. Hydrogen chloride converts it into trichlorogermane; nitric or sulphuric acids oxidize it to germanium dioxide and sulphur—

$$GeS \xrightarrow[350°C]{O_2} GeO_2 + SO_2$$
$$GeS \xrightarrow[150°C]{HCl} HGeCl_3 + H_2S$$

D. Germanium(II) halides

All of the dihalides are known; they show a most surprising variation in properties, GeF_2 and GeI_2 being the most stable.

1. *Germanium(II) fluoride*

This was first prepared in 1961 (Bartlett and Yu; 1961; Muetterties and Castle, 1961)—

$$GeF_4 + Ge \xrightarrow{150°-300°C} GeF_2$$

$$Ge + HF \xrightarrow{225°C} GeF_4 + GeF_2$$

It is a white crystalline solid, melting at 110°C, and may be distilled *in vacuo* without decomposition. Above 160°C it decomposes into the volatile tetra-fluoride and an orange–red solid, presumed to be a polymeric sub-fluoride. The crystal lattice of germanium(II) fluoride consists of a fluorine-bridged chain polymer in which parallel chains are cross-linked by weak fluorine bridges. The germanium co-ordination may be considered as a distorted trigonal bipyramidal arrangement of four fluorine atoms and one non-bonding electron pair (Trotter *et al.*, 1966). Germanium difluoride in a mass spectrometer produces $(GeF_2)_n^+$ ions ($n = 1$–4), the ionization potential of GeF_2 being 11.8 ± 0.1 eV. Thermodynamic values for the reaction $GeF_2(s) \rightarrow GeF_2(g)$ are: $\Delta H^0_{361} = 27.0$ kcal. mole^{-1}, $\Delta S^0_{361} = 46 \pm 3$ e.u. and $\Delta G^0_{361} = 10.4$ kcal. mole^{-1} (Zmbov *et al.*, 1968).

Germanium difluoride is rapidly hydrolysed by water and is evidently a weak electron acceptor forming a 1:1 complex with dimethyl sulphoxide (Muetterties, 1962). It is a strong reducing agent and reacts explosively with sulphur trioxide, although selenium tetrafluoride is reduced to selenium—

$$2GeF_2 + SeF_4 \rightarrow 2GeF_4 + Se$$

2. *Germanium(II) chloride*

Whereas germanium tetrachloride is stable at least to 1300°K, vapour-pressure studies in the presence of a tenfold excess of metallic germanium show that up to 570°K only germanium tetrachloride is present in the vapour phase. Between 570° and 950°K the slope of the pressure–temperature curve increases, and above 950°K it corresponds to twice the number of molecules in the gas phase. These observations are compatible with the reaction—

$$Ge(s) + GeCl_4(g) \rightleftharpoons 2GeCl_2(g)$$

for which $\Delta H^0 = 34.9$ kcal. mole^{-1} and $\Delta S^0 = 46.7$ e.u. (Sedgwick, 1965).

Earlier work by Dennis and Hunter (1929) demonstrated that passage of germanium tetrachloride over germanium at 430°C produced the dichloride as a white solid that deposited beyond the heated zone. It has also been reported, though without detail, that silver chloride and germanium at 700°C give germanium tetra- and di-chlorides (Lieser *et al.*, 1961).

Reports on the thermal stability of germanium dichloride vary widely. Dennis and Hunter (1929) found that, in an evacuated sealed system dissociation began at 75°C. Moulton and Miller (1956) reported that decomposition to a sub-chloride ($GeCl_{0.64}$) occurs even at $-20°C$, but Mironov and Gar (1965a) were unable to repeat this work. Other early work (Dennis *et al.*, 1926) also indicates that it is stable at least up to room temperature. Dissociation of trichlorogermane produces germanium dichloride—

$$HGeCl_3 \rightleftharpoons GeCl_2 + HCl$$

Germanium dichloride is described as a white crystalline solid, rapidly hydrolysed by water and slowly oxidized in air, but with no indication of oxychloride formation—

$$2GeCl_2 + O_2 \rightarrow GeO_2 + GeCl_4$$

It is fairly soluble in germanium tetrachloride, readily oxidized by halogens and is converted into the sulphide by hydrogen sulphide. With hydrogen chloride in ether it forms a yellow oily complex having the composition $HGeCl_3.2Et_2O$, which decomposes *in vacuo* to germanium dichloride.

3. *Germanium(II) bromide*

This compound has been obtained by the action of strong hydrobromic acid on germanium(II) hydroxide (Mironov and Gar, 1965a)—

$$Ge(OH)_2 + 2HBr \rightarrow GeBr_2 + 2H_2O$$

Its properties are closely related to those of the dichloride.

4. *Germanium(II) iodide*

This is best made by the reduction of germanium tetraiodide with hypophosphorous acid—

$$GeO_2 \xrightarrow{HI} GeI_4 \xrightarrow{H_3PO_2} GeI_2$$

Residual germanium tetraiodide may be sublimed from the product *in vacuo* at 100°C, the yellow solid residue being relatively pure germanium di-iodide. Other, less convenient, preparative methods start from germanium(II) compounds ($GeO + HI$; $GeCl_2 + KI$). It crystallizes with the cadmium iodide lattice, is insoluble in hydrocarbons, slightly soluble in ethers, and may be recrystallized from strong aqueous hydrogen iodide. In air it oxidizes

2*

to the dioxide and tetraiodide above 200°C, and thermal decomposition (270°–370°C) gives germanium and the tetraiodide, but with an inert carrier gas some sublimation of the di-iodide is also observed (Foster, 1950; Jolly and Latimer, 1952a). It reacts only slowly with cold water and is freely soluble in ethanol with discharge of the colour. Probably its most important reactions are those with organic halides, which are discussed later (Chapter 7). With hydrogen chloride in ether it gives a red oily liquid corresponding to $HGe(Cl)I_2 . 2Et_2O$ (Nefedov and Kolesnikov, 1966). Germanium di-iodide forms 1:1 complexes with many tertiary and secondary phosphines—

$$R_3P + GeI_2 \rightarrow R_3PGeI_2$$

The complexes, most of which are yellow crystalline solids, turn red on exposure to air, and molecular-weight measurements in benzene suggest some dissociation in solution. When heated *in vacuo*, complete dissociation occurs (King, 1963).

E. Germanium(II) phosphite

Reduction of germanium dioxide with the theoretical amount of hypo-phosphorous acid in excess of phosphorous acid gives the white crystalline phosphite (Everest, 1953)—

$$GeO_2 + H_3PO_2 \rightarrow GeHPO_3 + H_2O$$

It dissolves in strong or dilute halogen acids and decomposes without melting about 230°C.

F. Germanium(II) nitride

This was first obtained as a dark-brown powder by Johnson and Ridgely (1934) who decomposed polymeric germanium(II) imide (Johnson *et al.*, 1932)—

$$GeI_2 \xrightarrow{\ NH_3\ } (GeNH)_n + NH_4I$$

$$3GeNH \xrightarrow{\ 250°-300°C\ } Ge_3N_2 + NH_3$$

At higher temperatures, it decomposes to the elements; it has been reduced by hydrogen and is rapidly hydrolysed by caustic soda—

$$Ge_3N_2 \begin{cases} \xrightarrow[600°C]{H_2} 3Ge + 2NH_3 \\ \xrightarrow{850°C} 3Ge + N_2 \\ \xrightarrow{H_2O} 3Ge(OH)_2 + 2NH_3 \end{cases}$$

Germanium(II) nitride is also formed when monogermane is decomposed by active nitrogen (Storr *et al.*, 1962)—

$$3GeH_4 + N_2* \xrightarrow{100°-200°C} Ge_3N_2 + 6H_2$$

G. Germanites

Muetterties (1962) obtained trifluorogermanites, such as the caesium salt, as crystalline solids—

$$CsF + GeF_2 \rightarrow CsGeF_3$$

The $GeF_3{}^-$ ion is hydrolytically stable and the crystalline salt is not oxidized in the absence of moisture, but in solution oxygen is absorbed and with excess fluoride present the fluorogermanate, Cs_2GeF_6 is obtained. The reducing properties of germanites are well illustrated by the reaction with strong aqueous hydrofluoric acid—

$$CsF + CsGeF_3 + 2HF \rightarrow H_2 + Cs_2GeF_6$$

Trichlorogermanites have likewise been obtained from trichlorogermane (Tananaev *et al.*, 1964), and by the reduction of germanium tetrachloride with hypophosphorous acid (Poskozim, 1968)—

$$NH_4Cl + HGeCl_3 \longrightarrow NH_4GeCl_3 + HCl$$
$$GeCl_4 + H_3PO_2 \xrightarrow{CsCl} CsGeCl_3$$

The ammonium salt is stable to 110°C, and is not hygroscopic.

Another co-ordination complex of germanium(II) has been obtained by the reaction (Johnson *et al.*, 1966)—

$$Ph_4AsCl + HGeCl_3 \rightarrow [Ph_4As]^+[GeCl_3]^-$$

This colourless crystalline salt forms a further series of complexes with strong electron donors—

$$[Ph_4As]^+[GeCl_3]^- \xrightarrow{BX_3} [Ph_4As]^+[X_3BGeCl_3]^-$$

IV. GERMANIUM(IV) COMPOUNDS

A. Germanium(IV) oxide

This is usually obtained by the hydrolysis of germanium tetrachloride with aqueous ammonia and it exists in several forms.

1. *Vitreous GeO₂*

This is a true glass formed when the fused dioxide solidifies; it dissolves to the extent of 10 g litre^{-1} in cold water (Schwarz and Huf, 1931). The properties of liquid germanium dioxide have been studied (Mackenzie, 1958).

2. Hexagonal GeO_2

This is the common form produced by the hydrolysis of germanium tetrahalides. Its solubility in water shows a slow increase to a maximum of 4·6 g litre^{-1} after 10 days at room temperature, but in hot water it dissolves rapidly to the extent of 7·5–8·0 g litre^{-1} (Mueller, 1926; Zachariasen, 1928).

3. Tetragonal GeO_2

This form has a rutile type structure and is obtained by heating at about 380°C the residue obtained from the evaporation of an aqueous solution of germanium dioxide (Mueller and Blank, 1924). Its solubility is only 0·1–0·2 g litre^{-1} in hot or cold water (Murthy and Hill, 1965), and it is also virtually insoluble in hydrochloric and hydrofluoric acids, but dissolves slowly in 5M sodium hydroxide at 100°C. The radius ratio for the hexagonal and tetragonal forms is close to the value expected for a change from octahedral to tetrahedral co-ordination (Smith and Isaacs, 1964).

The hexagonal form of germanium dioxide dissolves readily in hydrofluoric acid in an exothermic reaction; it combines with bromine trifluoride, also giving germanium tetrafluoride, and reacts with carbon tetrachloride at 500°C (Dede and Russ, 1928; Emeléus and Woolf, 1950). Heats of formation of the three forms are given in Table II (Gross et al., 1966)—

$$\Delta H^0_{f\ 298}\ GeO_2\ (\text{hexagonal} \rightarrow \text{tetragonal}) = -6\cdot08 \pm 0\cdot34 \text{ kcal. mole}^{-1}$$

$$\Delta H^0_{f\ 298}\ GeO_2\ (\text{glass} \rightarrow \text{hexagonal}) = -3\cdot75 \pm 0\cdot14 \text{ kcal. mole}^{-1}$$

B. Germanic acids and their salts

There have been several investigations into the state of germanium dioxide in aqueous solution using titrimetric and ion-exchange techniques (Everest and Salmon, 1954, 1955), and the whole field of germanate chemistry has been surveyed by Ingri (1963). In the pH range 6·9–9·4 the predominant species is said to be the pentagermanate ion, $Ge_5O_{11}{}^{2-}$, and crystalline salts such as $K_2Ge_5O_{11}$ have been isolated from germanium dioxide solutions in 2M potassium chloride at pH 9·2 (Tchakirian and Carpéni, 1948). At both high and low pH values, depolymerization of pentagermanate ions to monomeric germanic acid occurs. In 0·03–0·07M sodium hydroxide it exists as the $HGeO_3{}^-$ ion (Gayer and Zajicek, 1964b), but there is no evidence of intermediate telomer ions such as $Ge_4O_9{}^{2-}$. In the presence of sulphate or phosphate at pH 7–9·5, the complex ions $[GeO_2(SO_4)]^{2-}$ and $[HGeO_2(PO_4)]^{2-}$ are formed. Ingri considers that the pentagermanate ion s more accurately represented by $[(OH)_3[Ge(OH)_4]_8]^{3-}$. Dissociation constants of metagermanic acid, H_2GeO_3, have been reported as $K_1 = 3\cdot2 \times 10^{-9}$ and $K_2 = 1\cdot9 \times 10^{-13}$ (Pugh, 1929), but the free acid has never been isolated. It condenses with hydroxy-compounds, such as glycerol and mannitol, to give esters that are much more acidic and may be titrated with caustic soda as monobasic acids.

Magnesium orthogermanate, Mg_2GeO_4 has been isolated as a white powder by treating ammonia solutions of germanium dioxide with magnesium oxide (Mueller, 1922). Unequivocal evidence for the existence of orthogermanic acid, $(HO)_4Ge$, in solution is lacking, but the ion $[Ge(OH)_6]^{2-}$ is found in the crystalline mineral stoettite, $FeGe(OH)_6$ (Strunz and Giglio, 1961). Sodium metagermanate, Na_2GeO_3, forms a white hygroscopic solid that absorbs carbon dioxide and is almost completely hydrolysed in aqueous solution.

$$GeO_2 + Na_2CO_3 \xrightarrow{\text{high temp.}} CO_2 + Na_2GeO_3$$

Various crystalline hydrates ($Na_2GeO_3 . 6H_2O$, $Na_2GeO_3 . 7H_2O$) have been isolated, and the systems Na_2GeO_3–GeO_2 and K_2GeO_3–GeO_2 show the existence of a range of condensed anions such as $Na_2Ge_2O_5$ and $Na_2Ge_4O_9$ (Schwarz and Lewinsohn, 1930; Schwarz and Heinrich, 1932). The crystal structure of $Na_4Ge_9O_{20}$ is built up of GeO_6 octahedra and GeO_4 tetrahedra, coupled together forming a three dimensional network (Ingri and Lundgren, 1963), whereas $Li_2Ge_2O_5$ has a layer structure, and is isostructural with $Li_2Si_2O_5$ (Modern and Wittman, 1965).

C. Germanium(IV) alkoxides

Alkyl orthogermanates, i.e., esters of orthogermanic acid, are made by the alcoholysis of germanium tetrachloride in the presence of a base, such as sodium alkoxide or pyridine, or they can be obtained by alcohol–exchange reactions. Glycols and catechols chelate to germanium—

$$GeCl_4 \xrightarrow[\text{py}]{ROH} Ge(OR)_4 \xrightarrow{R'OH} Ge(OR')_4$$

(Bradley et al., 1956; Mehrotra and Chandra, 1963; Pichet and Benoit, 1967). All examples so far reported are colourless liquids [$Ge(OEt)_4$ has b.p. 187°C] that are rapidly hydrolysed by water. Tetraethoxygermane was first prepared by Tabern et al. (1925), and equilibrium constants have been evaluated for the ethanolysis of germanium tetrachloride (Shevchenko and Kuzina, 1963)—

$$GeCl_4 + nEtOH \rightleftharpoons (EtO)_nGeCl_{4-n} + nHCl$$

$$K_1 = 9\cdot8 \times 10^{-2} \qquad K_3 = 2\cdot5 \times 10^{-5}$$
$$K_2 = 2\cdot5 \times 10^{-3} \qquad K_4 = 2\cdot3 \times 10^{-8}$$

Similarly the formation of tetramethoxygermane has been followed potentiometrically (Gut, 1964). The partial hydrolysis of tetraisopropoxy-germane has given a volatile germoxane together with solid polymeric products (Schwarz and Krauff, 1954)—

$$(Pr^iO)_4Ge \xrightarrow{<4H_2O} [(Pr^iO)_3Ge]_2O + [Ge_2O_3(OPr^i)_2]_n$$

Tetraethoxygermane treated with four equivalents of water in ethanol solution, forms a germanium dioxide gel which has considerable adsorptive power for benzene, carbon tetrachloride and ether (Laubengayer and Brandt, 1932a). A further versatile preparative method makes use of tri-alkyl orthoformates—

$$GeCl_4 + 4HC(OR)_3 \xrightarrow{AlCl_3} Ge(OR)_4 + 4RCl + 4HCO_2R$$

The reaction can be controlled by altering the stoicheiometry, so that quite high yields of the alkoxychlorogermanes [$ROGeCl_3$, $(RO)_2GeCl_2$ and $(RO)_3GeCl$] are obtained (Pike and Dewidar, 1964). Cyclic ethers are also cleaved by germanium tetrachloride (Lavigne et al., 1967)—

$$GeCl_4 + \underset{O}{\overset{CH_3}{\triangle}} \xrightarrow{MgBr_2} Cl_xGe(OCH_2CHClCH_3)_{4-x}$$

Alkoxygermanium halides are formed in redistribution reactions between tetra-alkoxygermanes and germanium tetrahalides, and, for tetramethoxy-germane, there are large deviations from random distribution at equilibrium, the mixed species being more abundant (Bottei and Kuzma, 1968).

A germanium phosphate [$(HPO_4)_2Ge.H_2O$] and arsenate have been described (Avduevskaya et al., 1964; Avduevskaya and Tananaev, 1965), and the highly stable trifluoroacetate, $Ge(OCOCF_3)_4$, has been isolated from the reaction (Sartori and Weidenbruch, 1967)—

$$GeCl_4 + (CF_3CO_2)_2Hg \rightarrow Ge(OCOCF_3)_4$$

D. Germanium(IV) sulphide

This is precipitated as a white solid when hydrogen sulphide is passed into strongly acid solutions (e.g., 6M sulphuric acid) of any germanium(IV) compound (Johnson and Dennis, 1925). It is also formed when the dioxide is heated with sulphur, and by the direct combination of germanium and sulphur at 1100°C and high pressure. The latter method gives a tetragonal form as GeS_2 (Prewitt and Young, 1965).

Germanium disulphide melts about 800°C to a dark liquid, and vaporizes about 850°C. It dissolves readily in hot caustic soda solution and in ammonium hydroxide. It is unaffected by concentrated hydrochloric acid, but nitric acid converts it into the dioxide.

E. Alkylthiogermanes

Compounds of the type $(RS)_4Ge$ are readily prepared in a mildly exothermic reaction—

$$GeCl_4 + 4RSH \xrightarrow{\text{NH}_3,\ \text{benzene}} (RS)_4Ge + 4NH_4Cl$$

Sterically hindered groups react to give trisubstituted derivatives with ammonolysis of the fourth halogen—

$$GeCl_4 + 3Bu^tSH + 5NH_3 \rightarrow (Bu^tS)_3GeNH_2 + 4NH_4Cl$$

They are mostly colourless viscous liquids. In the presence of a proton source (e.g., p-toluenesulphonic acid) mercaptans react with alkoxygermanes, the equilibrium lying well to the right (Mehrotra et al., 1967a)—

$$(Pr^iO)_4Ge + 4RSH \rightleftharpoons (RS)_4Ge + 4Pr^iOH$$

F. Germanium(IV) halides

1. Germanium tetrafluoride

Barium hexafluorogermanate, which separates as a white granular precipitate when a strong aqueous solution of barium chloride is added to germanium dioxide dissolved in 47% hydrofluoric acid, decomposes in a quartz tube at 700°C steadily evolving germanium tetrafluoride (Hoffman and Gutowsky, 1953). Germanium tetrafluoride is a colourless gas that fumes in air and has a garlic-like smell. It only attacks glass in the presence of moisture. It boils at −36·5°C; the triple point is at −15°C and 3032 mm. Hydrolysis is both rapid and exothermic—

$$3GeF_4 + 2H_2O = GeO_2 + 2H_2GeF_6$$

Fluorogermanic acid formed in the hydrolysis can be precipitated as the potassium salt. Germanium tetrafluoride is a strong electron acceptor and forms many complexes with donor molecules; these are discussed in Section G. It is more reactive than silicon tetrafluoride towards halide exchange. For example, with a variety of anhydrous metal chlorides over the range 200°–600°C it gives only germanium tetrachloride with no evidence for chlorofluorogermanes. This contrasts with silicon, which under similar conditions gives the full range of chlorofluorosilanes (Schumb and Breck, 1952). Heat capacities and entropies of germanium(IV) halides have been reported (Jolly and Latimer, 1952a).

2. Germanium tetrachloride

This compound is available commercially, but it may be prepared by the action of concentrated hydrochloric acid on germanium dioxide or from germanium and chlorine (Foster *et al.*, 1946). It is a colourless liquid (b.p. 83·1°C, m.p. −49·5°C, density 1·88) that fumes in air and is quite rapidly hydrolysed by water. It dissolves freely in most organic solvents, is only sparingly soluble in concentrated hydrochloric acid (Allison and Mueller, 1932) and is unaffected by concentrated sulphuric acid. Partial hydrolysis at −78°C in ether, chloroform or pentane solution does not produce a volatile oxychloride, but rather white benzene-insoluble polymeric solids approximating in composition to $Ge_2O_3Cl_2$ (Schumb and Smyth, 1955). The reaction between germanium tetrachloride and oxygen at 950°C is said to give low yields of a liquid oxyhalide, $Cl_3GeOGeCl_3$ boiling at 56°C and 13 mm (Schwarz *et al.*, 1931).

3. *Germanium tetrabromide*

This can be obtained in high yield by the action of hydrobromic acid on germanium dioxide, from which it may be completely extracted with ether (Laubengayer and Brandt, 1932b; Ladenbauer *et al.*, 1955). The action of bromine on finely divided germanium has also been used as a preparative method (Brauer, 1963). Germanium tetrabromide is a colourless crystalline solid (m.p. 26·0°C; b.p. 186°C) that fumes in moist air, is rapidly hydrolysed by water and is freely soluble in organic solvents.

4. *Germanium tetraiodide*

This compound is best made from germanium dioxide and strong aqueous hydriodic acid, from which solution it separates in high yield. (Laubengayer and Brandt, 1932b; Foster and Williston, 1946). Like the other tetrahalides it can also be made from the metal and halogen (Dennis and Hance, 1922). The reaction begins at 220°C and is vigorous at 560°C. Germanium tetraiodide forms orange cubic crystals (m.p. 146°C) that sublime readily *in vacuo* at about 80°C, and this is the most satisfactory method of purification, although it may also be crystallized from chloroform. It decomposes above 440°C, dissolves in most organic solvents and is slowly hydrolysed by water giving a clear colourless acidic solution. It is unaffected by concentrated sulphuric acid over 24 hr at room temperature, but at 80°C iodine is slowly liberated. In concentrated nitric acid it turns black as nitric oxide is evolved. There is evidence from Raman spectra that halide exchange occurs in mixtures of germanium tetrahalides giving, for example, $GeBr_2ClI$ and $GeBrCl_2I$ (Cerf and Delhaye, 1964).

5. *Sub- and catenated-germanium halides*

Whereas silicon forms a range of catenated chlorides at least up to Si_6Cl_{14}, the only fully characterized germanium analogue is Ge_2Cl_6, which is best made by passing germanium tetrachloride at 0.1 mm through a microwave discharge tube. In this way, digermanium hexachloride may be collected in a trap at −18°C at a rate of 250 mg h⁻¹. It is a colourless crystalline solid (m.p. 40°–42°C) that sublimes readily at room temperature (Shriver and Jolly, 1958). No chemical reactions have so far been reported, but it is also formed in low yield in the reaction between germanium and germanium tetrachloride, together with ill defined sub-chlorides (Schwarz and Baronetzky, 1954). If germanium tetrachloride in a hydrogen stream is pyrolysed at 900°–1000°C, there is evidence of coloured sub-chlorides that do not show X-ray lines corresponding to either the metal or the dichloride. Other reports of sub-chlorides and sub-fluorides have already been mentioned under the germanium dihalides.

G. Octahedral complex ions

The acceptor power of the group IVb metal tetrahalides increases in the order Si < Ge < Sn. Thus silicon forms only the SiF_6^{2-} ion, whereas both hexafluoro- and hexachloro-germanates have been obtained, and all four hexaco-ordinate tin ions are known (SnF_6^{2-}, $SnCl_6^{2-}$, $SnBr_2^{6-}$ and SnI_6^{2-}). In K_2GeF_6 the fluoride ions are octahedrally disposed about germanium (Hoard and Vincent, 1942); it has a trigonal structure that becomes hexagonal at 410°C or on recrystallizing from bromine trifluoride (Brown et al., 1967). Vibrational frequencies and force constants have been determined (Begun and Rutenberg, 1967).

If germanium-tetrafluoride or -tetrachloride is added to a suspension of caesium chloride in liquid sulphur dioxide and the mixture maintained at 22°C under pressure for several days, the complex halide is readily isolated—

$$4CsCl + 3GeF_4 \xrightarrow[22°C]{SO_2} 2Cs_2GeF_6 + GeCl_4$$

In basic solution, the rate of hydrolysis of the hexafluorogermanate ion is first order with respect to GeF_6^{2-} and zero order with respect to hydroxide ion. In unbuffered aqueous solution the hydrolysis is autocatalytic and even at pH 4 almost all of the fluoride is transformed into F^- in a reaction which may be written as (Gebala and Jones, 1967)—

$$GeF_6^{2-} + (2 + x)H_2O \rightarrow GeO_2xH_2O + 6F^- + 4H^+$$

The complex chloride Cs_2GeCl_6 has been obtained as a white precipitate from germanium tetrachloride and caesium chloride in aqueous alcoholic hydrochloric acid (Klanberg, 1963). Its dissociation pressure over the range 140°–190°C is given by—

$$\log_{10} P = 14{\cdot}87 - \frac{5751}{T}$$

Germanium tetrafluoride, being a strong electron acceptor, forms both 1:1 and 1:2 complexes with electron-donor molecules. The tetrachloride and tetrabromide form similar addition complexes, but these are more readily dissociated (Ferguson et al., 1959; Muetterties, 1960; Lappert, 1962; Beattie et al., 1963; Frazer et al., 1964; Langer and Blut, 1966; Miller and Onyszchuk, 1967). Complete vibrational assignments have been made for the $GeCl_6^{2-}$ ion, together with approximate calculations of the vibrational frequencies for co-ordination complexes of the group IV metal tetrahalides (Beattie et al., 1964, 1967). Comparative studies indicate that SiF_4 and GeF_4 are weaker acceptors than SnF_4, TiF_4 and MoF_4, and that germanium tetrahalides are rather stronger than silicon analogues. Complexes with nitrogen, phosphorus, oxygen and sulphur donors have been described.

If germanium tetrafluoride is passed into a solution of an amine in ether or benzene, the complexes separate immediately and may often be recrystallized from ether or acetonitrile. Trimethylamine and acetonitrile have yielded crystalline 1:1 and 1:2 complexes [Me_3NGeF_4, $(Me_3N)_2GeF_4$, $MeCNGeF_4$, $(MeCN)_2GeF_4$], whereas other amines have only given 1:2 complexes [py, NH_3, N_2H_4, piperidine]. Ethylenediamine, NN'tetramethylethylenediamine, bipyridyl and o-phenanthroline all give 1:1 complexes having a low dissociation pressure in which the amine is almost certainly chelating. All of the nitrogen complexes are white solids, decomposed by water, alcohols and other protic solvents, and are almost, if not completely, dissociated in the vapour state. Heats of reaction (which relate to the donor–acceptor bond strengths) of pyridine and isoquinoline with germanium tetrahalides decrease in the order $GeF_4 > GeCl_4 > GeBr_4$ for the 1:2 complexes.

Trimethylphosphine and even phosphine form 1:1 complexes with germanium tetrafluoride; these are white solids that are said to sublime with some dissociation at 25°C. Oxygen donors, such as dimethyl sulphoxide, trimethylamine oxide, dimethylformamide, acetoxime, ketones, ethers and methanol, also yield 1:2 complexes, although those involving ketones, ethers and methanol are of low stability. Sulphur donors form both 1:1 and 1:2 complexes [H_2SGeF_4, $(H_2S)_2GeF_4$, $(Me_2S)_2GeF_4$]. The reactivity of MF_4 complexes varies greatly: $(Me_2SO)_2TiF_4$ can even be recrystallized from water.

The structure of these complexes is of some interest. The 1:2 adducts are undoubtedly octahedral with the possibility of cis–trans forms. In several tin complexes of the type (amine)$_2$SnF$_4$, the ^{19}F resonance is split into two triplets of equal intensity, which is compatible with a cis configuration. However the ^{19}F spectra of silicon tetrafluoride– and germanium tetrafluoride–amine complexes show only a single fluorine resonance. This may indicate a trans structure, but it could be the result of rapid ligand exchange. In fact carbonyl complexes such as (2,6-dimethyl-γ-pyrone)$_2$GeF$_4$ produce a broad singlet ^{19}F resonance at room temperature which splits into the two triplets of an A$_2$X$_2$ spectrum on cooling; hence the *cis* isomer is present at the lower temperature. This and similar germanium tetra-fluoride complexes may be recovered unchanged from aqueous solution (Adley, *et al.*, 1968).

The 1:1 complexes are mostly not sharp melting; they have a low solubility in both non-polar and polar (non-protic) solvents, and it seems most unlikely that they are monomeric pentaco-ordinate complexes. The specific conductance of Me$_3$NGeCl$_4$ (m.p. 75°–76°C) in acrylonitrile is low compared to the ionic compound [Me$_3$NSiH$_3$]I. The indications are that the 1:1 complexes are polymeric fluorine- or chlorine-bridged complexes in which the metal is effectively 6 co-ordinate. However germanium tetra-chloride and silver perchlorate combine in the presence of trimethylamine forming the salt [(Me$_3$N)$_2$GeCl$_3$]ClO$_4$. Its vibrational spectrum is compatible with a monomeric 5 co-ordinate ion of D_{3h} symmetry (Beattie *et al.*, 1968). Some examples of the use of far i.r. spectroscopy in deducing the probable stereochemistry of 6 co-ordinate complexes have been given by Beattie *et al.* (1963). Tetraphenylarsonium chloride and germanium dioxide in hydrofluoric acid give a crystalline salt having the composition Ph$_3$AsGeF$_5$; this may be an example of a pentaco-ordinate GeF$_5^-$ ion (Clark and Dixon, 1967).

Germanium tetrafluoride forms a number of complexes with *o*-phen-anthrolineiron(II) cyanide. The 1:1 complex is evidently a dimer and may result from the nitrogen atoms of the cyanide groups donating electrons to germanium—

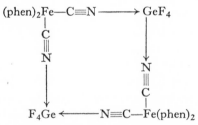

Related complexes of different stoicheiometry, (phen)$_2$Fe(CN)$_2$.2GeF$_4$ and (phen)$_2$Fe(CN)$_2$.3GeF$_4$, probably involve germanium–fluorine–germanium bridging groups (Shriver, 1966).

Among the most interesting complexes are those with xenon hexafluoride (Pullen and Cady, 1967). Germanium tetrafluoride and xenon hexafluoride form, at room temperature, a white solid having the composition (XeF$_6$)$_4$GeF$_4$ that loses XeF$_6$ at 0°C *in vacuo*, forming a white crystalline complex, (XeF$_6$)$_2$GeF$_4$. This 2:1 complex appears to be more stable than the 1:1 adduct $(XeF_6)GeF_4$, which is also a white crystalline solid and is obtained if an excess of germanium tetrafluoride is mixed with xenon hexafluoride. The 1:1 complex will sublime above 50°C. Tin tetrafluoride gives similar complexes, but silicon tetrafluoride is evidently unreactive; this is in line with the relative acceptor properties discussed earlier.

Phenol complexes of germanium have received some study (Drozdova *et al.*, 1962; Yoder and Zuckerman, 1967). Thus catechol and germanium tetrachloride in the presence of pyridine yield an octahedral complex, whereas 2,2′-dihydroxybiphenyl and 1,8-dihydroxynaphthalene form tetra-co-ordinated spiro-esters free from associated base—

The trisoxalatogermanate ion has been isolated and separated into its optical isomers (Moeller and Nielsen, 1953). The potassium salt, $K_2Ge(C_2O_4)_3 \cdot H_2O$, forms white crystals that decompose at 180°C, with loss of carbon dioxide. Other complex oxalates, such as $[Co(NH_3)_6][Ge(C_2O_4)_3]_3$, show a similar thermal stability (Kurnevich and Shagisultanova, 1964; Arvedson and Larsen, 1966). Mandelic acid forms a monoclinic complex with germanium dioxide in which there are two germanium atoms, four mandelic acid residues and four water molecules per unit cell (Clark, 1961; Sterling, 1967). Aqueous solutions of germanium dioxide form 1:2 complexes with 8-hydroxyquinoline, $(HO)_2Ge(oxine)_2$, which are only slowly and incompletely precipitated (Clero, 1966; Tsau *et al.*, 1967). Acetylacetone and other β-diketones react vigorously with germanium tetrachloride in anhydrous solvents with evolution of hydrogen chloride and quantitative formation of the sparingly soluble octahedral complex $(acac)_2GeCl_2$. The analogous complex formed by propionylacetone is more soluble and is monomeric in benzene. Germanium tetrabromide is much more sluggish in its reaction with acetylacetone (Morgan and Drew, 1924). The p.m.r. spectrum of the acetylacetonate strongly suggests a *cis* configuration (Smith and Wilkins, 1966); its dipole moment in chloroform solution is 7·4D (Osipov *et al.*, 1966).

Trisacetylacetonates of germanium have not been obtained by the direct reaction, but cupric acetylacetonate and germanium tetrachloride give a mixture of products when heated in chloroform from which the colourless cationic tris complex $[(acac)_3Ge]^+[CuCl_2]^-$, may be isolated. The $FeCl_4^-$ salt has been obtained by the reaction (Mueller and Heinrich, 1962; Org and Prince, 1965)—

$$(acac)_2GeCl_2 + (MeCO)_2CH_2 + FeCl_3 \rightarrow [(acac)_3Ge]^+[FeCl_4]^-$$

A phthalocyanine complex of germanium has been obtained both from phthalocyanine and germanium tetrachloride in quinoline at 240°C (Joyner and Kenney, 1960) and from 2-cyanobenzamide and germanium tetrachloride in refluxing 1-chloronaphthalene (Joyner et al., 1962). A related tetradentate ligand, hemiporphyrazine, forms a complex of the phthalocyanine type (Esposito et al., 1967).

(2)

Phthalocyaninegermanium dichloride is blue–green by transmitted light and shows the high thermal stability (sublimes in vacuo at 450°C) common to so many metal phthalocyanines. The high stability of the ring system requires that the four germanium—nitrogen bonds lie in a plane with the two chlorine atoms mutually trans (2). This complex is of special interest, since few metals as electropositive as germanium have so far yielded phthalocyanine derivatives. The chlorine atoms may be substituted without breakdown of the complex. For example, hydrolysis proceeds with difficulty, giving a blue complex, probably the most stable dihydroxy derivative of germanium. On heating it dehydrates completely forming an oxygen-bridged polymer.

H. Inorganic germanium hydrides

The group IV metal hydrides have been discussed in two monographs (Stone, 1962; Mackay, 1966). For all of the group IV elements, the thermal stability of their hydrides decreases progressively from carbon to lead. However their sensitivity to oxygen is not progressive; germanes being less prone to atmospheric oxidation than either silanes or stannanes. Two tin hydrides (SnH_4, Sn_2H_6) have been characterized, but the only binary lead hydride is PbH_4, and this is desperately unstable thermally. A further anomaly in the series is the resistance of germanes to base hydrolysis, which contrasts with analogous silanes and, to a lesser extent, stannanes. For example, ordinary glass vessels provide sufficient base to bring about the hydrolysis of silane, whereas germane is only slowly hydrolysed by quite strong aqueous base. This difference is most reasonably attributed to the greater polarity of the silicon—hydrogen bond ($\overset{\delta+}{Si}$—$\overset{\delta-}{H}$) which, combined with the availability of empty d orbitals on silicon, provides a low activation energy path for attack by nucleophiles.

1. *Binary hydrides*

Of the group IV metals, silicon has yielded the greatest number of binary hydrides. Thus hydrolysis of magnesium silicide with phosphoric acid has led to the chromatographic detection of 21 hydrides with formulae up to Si_8H_{18}, and the occurrence of isomeric polysilanes such as n-Si_5H_{12} and iso-Si_5H_{12} has been established (Phillips and Timms, 1963). The silanes are less stable thermally than the corresponding alkanes (at about 500°C they all decompose to silicon and hydrogen) and all except monosilane ignite spontaneously in air.

Germanium forms a similar series of hydrides. The earliest systematic studies made use of the acid hydrolysis of magnesium germanide as the source of mono-, di-, tri-, tetra- and penta-germanes (Dennis *et al.*, 1924; Amberger, 1959). Monogermane has also been obtained by the reduction of germanium tetrachloride with metal hydrides [$LiAlH_4$ (Finholt *et al.*, 1947); KBH_4 (Macklen, 1959; Jolly, 1961) and $Li(Bu^tO)_3AlH$ (Sujishi and Keith, 1958)]. The simplest method for preparing monogermane is the slow addition of a mixture of 3 moles of sodium borohydride and 1 mole of germanium dioxide in sodium hydroxide solution to dilute acetic acid. This gives monogermane in about 75 % yield, together with some 9 % of digermane and detectable amounts of higher germanes up to Ge_9H_{20}, including three isomeric pentagermanes (Drake and Jolly, 1962b, 1963). Lithium aluminium hydride is a less satisfactory reducing agent, giving germane in less than 40 % yield. A yellow–orange solid, probably $(GeH_2)_x$, separates as the reaction proceeds, and this may be formed by hydrogen chloride elimination from trichlorogermane—

$$GeCl_4 \xrightarrow{LiAlH_4} HGeCl_3 \longrightarrow GeCl_2 \xrightarrow{LiAlH_4} (GeH_2)_x$$

Sodium borohydride in anhydrous tetrahydrofuran reacts unexpectedly with germanium tetrachloride to give only low yields of GeH_4, but on hydrolysis much additional germane is formed. This suggests that a germanium borohydride, $H_n Ge(BH_4)_{4-n}$, is first formed and is subsequently hydrolysed to GeH_4, hydrogen and boric acid (Macklen, 1959).

The higher germanes are probably best obtained by circulating monogermane through a silent electric discharge; Ge_2H_6 may be condensed from the mixture at $-130°C$, Ge_3H_8 and the higher germanes at $-95°C$. The pure deuterogermanes, GeD_4, Ge_2D_6 and Ge_3D_8 have been obtained by the action of deuterium chloride in deuterium oxide on magnesium germanide (Zeltmann and Fitzgibbon, 1954). Physical properties of the first three members of the homologous series are given in Table XIII.

TABLE XIII

Germanes and deuterogermanes

Compound	M.p., °C	B.p. at 760 mm, °C	ΔH_{vap}, kcal. mole^{-1}
GeH_4	−165	−88·5	3·608
Ge_2H_6	−109	29(31)	5.99
Ge_3H_8	−105·6	110·5	—
GeD_4	−166·2	−89·2	3·744
Ge_2D_6	−107·9	28·4	6·483
Ge_3D_8	−100·3	110·5	7·876

A mixture of 23% monogermane and oxygen at an initial pressure of $\frac{1}{3}$ atm reacts slowly at 160°C. At higher temperatures ($>330°C$), mixtures of monogermane and oxygen explode. The oxidation is a chain reaction, strongly catalysed by the solid reaction product, H_2GeO_3 (Eméleus and Gardner, 1938)—

$$GeH_4 + 2O_2 \rightarrow H_2GeO_3 + H_2O$$

Digermane is rapidly oxidized at 100°C, and the higher germanes are so sensitive to oxygen that they can only be manipulated *in vacuo* or in an inert atmosphere (Gokhale, *et al.*, 1965).

The rate of thermal decomposition of monogermane becomes appreciable above 280°C and the reaction is evidently heterogeneous, the rate being proportional to $(P_{GeH_4})^{1/3}$ over the range $283°-374°C$. From the method of purification, the monogermane used in this kinetic study was probably contaminated with digermane (Hogness and Johnson, 1932). A more recent kinetic study combined with observations on deuterium exchange suggests that both homogeneous and heterogeneous decomposition processes operate simultaneously. The homogeneous process is most probably represented by—

$$GeH_4 \rightarrow GeH_2 + H_2$$

At high germane pressures (~400 mm) the decomposition is first order, but the order decreases to zero as the pressure is decreased (Tamaru et al., 1955; Fensham et al., 1955). The pyrolysis of digermane is also homogeneous and first order over a large part of its course. Between 195° and 222°C, the overall decomposition may be represented by —

$$Ge_2H_6 \rightarrow 1 \cdot 8GeH_4 + 0 \cdot 49H_2 + 0 \cdot 82GeH_{0 \cdot 3}$$

The dark solid that forms as the pyrolysis proceeds is probably a mixture of germanium and a polymeric sub-hydride of germanium. The activation energy of this radical reaction is $33 \cdot 7$ kcal. mole^{-1} (Emeléus and Jellinek, 1944) and compares with 50 kcal. mole^{-1} for the pyrolysis of disilane. If the primary step is $Ge_2H_6 \rightarrow 2GeH_3$ rather than $Ge_2H_6 \rightarrow GeH_4 + GeH_2$, then the activation energy may approximate to the germanium—germanium bond energy.

2. Polymeric germanium hydrides

Two ill-defined polymeric hydrides of germanium (GeH and GeH$_2$) have been described on which there is no structural information; quite probably they involve both metal—metal and hydrogen-bridge bonding. A yellow or orange solid approximating in composition to (GeH)$_n$ is obtained as a by product when a mixture of sodium germanate and a large excess of sodium borohydride is added to sulphuric acid (Drake and Jolly, 1963). It decomposes with a puff in air and explodes about 165°C. It is unaffected by dilute acid or alkali, but is decomposed by strong acid. Its i.r. spectrum shows a single germanium–hydrogen stretching frequency at $4 \cdot 9$ μm and it is evidently X-ray amorphous. Probably the same hydride is formed by the hydrolysis of sodium germanide (Dennis and Skow, 1930) or by the addition of ammonium bromide to sodium germanide in anhydrous ammonia (Kraus and Carney, 1934). It is oxidized by chlorine and by acid permanganate.

A report that another polymeric hydride of composition (GeH$_2$)$_n$ is obtained as a yellow solid by the hydrolysis of calcium germanide, CaGe, with hydrogen chloride has been disproved: the composition varies over the range GeH$_{0 \cdot 9}$ to GeH$_{1 \cdot 2}$ (Royen and Rocktaeschel, 1966).

Germylsodium and bromobenzene apparently undergo the following unexpected reaction—

$$H_3GeNa + PhBr \xrightarrow{\text{NH}_3} C_6H_6 + NaBr + GeH_2$$

The germanium hydride is described as soluble in liquid ammonia, from which it crystallizes as a white solid, possibly a solvate, unstable above

−33°C. Its formation may be the result of halogen–metal exchange, giving H_3GeBr, followed by dehydrobromination (Glarum and Kraus, 1950)—

$$GeH_2 \xrightarrow{\;-33°\;} GeH_4 + (GeH)_n$$

The action of a silent electric discharge on monogermane described earlier gives in addition to the higher germanes a yellow polymeric hydride.

I. Inorganic derivatives of mono-, di- and tri-germane

1. *Alkali metal derivatives*

When monogermane is added to sodium in liquid ammonia at −63·5°C the conductivity falls to a minimum when the GeH_4–Na ratio is about 0·6. The initial blue colour is discharged giving a pale-yellow solution and a yellow–green precipitate. This conductivity behaviour, coupled with measurements of hydrogen evolved, has been interpreted in terms of the initial formation of a disodio-derivative (Eméleus and Mackay, 1961)—

$$GeH_4 + 2Na \rightarrow Na_2GeH_2 + H_2$$
$$Na_2GeH_2 + GeH_4 \rightarrow 2NaGeH_3$$
$$Na_2GeH_2 + NH_3 \rightarrow NaGeH_3 + NaNH_2$$

Similar studies on digermane provide evidence of two competing reactions—

$$Ge_2H_6 \xrightarrow[NH_3]{Na} NaGeH_3 + Na_2Ge_2H_4 + H_2$$

The colour changes to pale green when the Ge_2H_6–Na ratio is 0·5. With trigermane, a colour change to dark red is observed when the Ge_3H_8–Na ratio is 0·25. The formation of germyl-alkali metal derivatives in liquid ammonia is complicated by slow solvolysis of the product. If potassium amide in liquid ammonia is used, the main reaction is the formation of germyl potassium, but some solvolysis is indicated since hydrogen is slowly evolved (Glarum and Kraus, 1950). A further study of the reaction between germane and sodium in liquid ammonia has produced analytical evidence for the formation of germanium imide, which separates as a white solid, $Ge(NH)_2(NH_3)_n$. The formation of this compound may be the result of successive attack by amide ion on germane with elimination of ammonia from the germylamine (Rustad and Jolly, 1967)—

$$GeH_4 + 2e \rightarrow GeH_3^- + H^-$$
$$NH_3 + H^- \rightarrow NH_2^- + H_2$$
$$GeH_4 + NH_2^- \rightarrow H_3GeNH_2 + H^-$$

Solutions of sodium or potassium in hexamethylphosphotriamide, $(Me_2N)_3PO$, produce germyl-sodium or -potassium (Cradock *et al.*,

1967b), and an even simpler preparative method consists in stirring finely powdered potassium hydroxide with germane in monoglyme solution.

Germylsodium is a white solid, insoluble in ethylamine; it may be crystallized from liquid ammonia as the ammoniate, $NaGeH_3.6NH_3$. Under reduced pressure, the solvating ammonia is lost in stages (Kraus and Carney, 1934). Both germylsodium and germylpotassium decompose above 100°C with loss of all the hydrogen—

$$KGeH_3 \rightarrow KGe + 1\tfrac{1}{2}H_2$$

Germyl-lithium diammoniate is obtained as a pale-grey solid, stable at room temperature (Amberger and Boeters, 1963)—

$$GeH_4 + Li \xrightarrow{\text{NH}_3} H_3GeLi.2NH_3$$

At room temperature germylsodium decomposes with loss of hydrogen and formation of the shock-sensitive polymeric germanium hydride, $(GeH)_x$. Germylsodium reacts with oxygen without gas evolution to give a white solid, probably $NaGeO_2OH$, and electrolysis in liquid ammonia gives germane and nitrogen at the anode (Teal and Kraus, 1950).

2. Germyl halides and pseudohalides

All three types (H_3GeX, H_2GeX_2, $HGeX_3$) are known, and are, in general, of lower thermal stability than the corresponding silyl compounds. This may be attributable to differences in the polarity of silicon–hydrogen and germanium–hydrogen bonds. The dipole moment of H_3GeCl (2·12 D) is considerably greater than that of H_3SiCl (1·30 D) (Mays and Dailey, 1952) and is compatible with a greater polarity of the germanium—chlorine bond due to less effective π-bonding than with silicon. A further interesting comparison between silyl and germyl compounds is the much lower stability of bisgermyloxide, $(H_3Ge)_2O$, and trisgermylamine, $(H_3Ge)_3N$. Anharmonic constants and thermodynamic functions have been derived for the monohalides (Nagarajan, 1964).

(a) *H_3GeX and H_2GeX_2 compounds.* Chlorination of germane by hydrogen chloride in the presence of aluminium chloride proceeds smoothly at room temperature, giving principally monochlorogermane together with dichloro-germane—

$$GeH_4 + HCl \xrightarrow{\text{AlCl}_3} H_3GeCl + H_2GeCl_2$$

Hydrogen bromide reacts in the same way, but rather more vigorously. All four compounds are colourless liquids with a nauseating smell. Hydrogen iodide, in the presence of aluminium iodide, reacts much more vigorously with germane giving hydrogen, germanium di-iodide and other products (Dennis and Judy, 1929). Mono- and di-iodogermane may however be obtained by the action of iodine on monogermane at room temperature, or by halide exchange reactions. Di-iodogermane is a colourless crystalline solid melting at 46°C, whereas germyl fluoride is unstable at 25°C—

$$H_3GeBr \underset{\overset{AgF}{\longrightarrow} H_3GeF \overset{25°C}{\longrightarrow} GeH_4 + H_2GeF_2}{\overset{HI}{\longrightarrow} H_3GeI}$$

Exchange reactions with silyl halides are rapid at room temperature, and equilibrium constants have been evaluated from the proton resonances (Cradock and Ebsworth, 1967)—

$$H_3GeX + H_3SiY \rightleftharpoons H_3GeY + H_3SiX \quad (X = Cl, Br, I)$$

With the exception of difluorogermane, the H_2GeX_2 compounds appear less thermally stable than H_3GeX. Thermal decomposition of dichlorogermane is complex—

$$H_2GeCl_2 \rightarrow Ge + GeCl_2 + GeCl_4 + H_3GeCl + HCl + H_2$$

By contrast, germyl chloride gives simpler products, and is evidently hydrolysed to germanium(II) oxide

$$2H_3GeCl \overset{25°C}{\longrightarrow} GeH_4 + 2HCl + Ge$$

A whole range of germyl pseudohalides have been obtained by the reaction (Srivastava et al., 1962; Srivastava and Onyszchuk, 1963)—

$$H_3GeBr + AgX \rightarrow H_3GeX + AgBr \quad (X = CN, CNO, SCN, OAc)$$

Germyl cyanide is too unstable at 25°C for effective purification, but its i.r. spectrum shows that it is mainly the cyanide rather than the isocyanide (Goldfarb and Zafonte, 1964). Its microwave spectrum has given the following bond lengths: C—N, 1·155; Ge—C, 1·919; Ge—H, 1·529 Å (Varma and Buckton, 1967). Both the cyanate and thiocyanate reactions yield the iso derivatives, H_3GeNCO and H_3GeNCS. Spectroscopic examination of germyl isothiocyanate and its deuteride suggest that the Ge–N–C–S group is non-linear (Davidson et al., 1967). Acetoxygermane has a vapour pressure of 9·7 mm at 0°C, and in general these pseudohalides decompose about 50°C with the formation of a polymeric germanium hydride—

$$H_3GeX \rightarrow (H_2Ge)_n + HX$$

Digermane has yielded a variety of monohalides by low-temperature reactions, which are necessary because of the thermal instability of digermyl halides (only the chloride, Ge_2H_5Cl, is reasonably stable at room temperature)—

$$Ge_2H_6 + I_2 \xrightarrow{-63°} H_3GeGeH_2I + HI$$

The iodide, the least stable of the monohalides, has been used to make the chloride and bromide—

$$H_3GeGeH_2I + AgX \rightarrow H_3GeH_2X + AgI \quad (X = Cl, Br)$$

Other reactions and preparative methods are illustrated below—

$$Ge_2H_6 + AgX \rightarrow H_3GeGeH_2X + Ag + \tfrac{1}{2}H_2$$

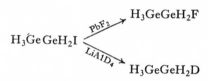

On thermal decomposition they give germane, digermane and various halogermanes together with involatile orange solids (Mackay and Roebuck, 1964; Mackay *et al.*, 1966).

Reaction of trigermane with iodine has given a highly unstable iodide, Ge_3H_7I, which, by *in situ* reduction with $LiAlD_4$, is converted into the monodeuteride, believed to be $(H_3Ge)_2GeHD$.

(b) *HGeX$_3$ compounds.* Dry hydrogen chloride combines rapidly with germanium at 480°–500°C to yield a liquid mixture consisting of some 70% trichlorogermane and 30% germanium tetrachloride. A pure product is more readily obtained by the addition of hydrogen chloride to germanium dichloride—

$$Ge + GeCl_4 \rightarrow GeCl_2 \xrightarrow[50°C]{HCl} HGeCl_3$$

Trichlorogermane (b.p. extrapolated 75·2°C; m.p. −71°C) is oxidized by iodine with liberation of hydrogen iodide, and decomposes rapidly above 140°C—

$$2HGeCl_3 \rightarrow 2HCl + 2GeCl_2 \xrightarrow{140°C} GeCl_4 + Ge$$

It combines exothermically with oxygen at 0°C, but the course of the reaction is not clear; germanium tetrachloride (but not hydrogen chloride) is formed together with an unstable viscous oil. Hydrolysis of trichlorogermane gives $Ge(OH)_2$ as an orange solid (Dennis *et al.*, 1926). There is much evidence that trichlorogermane is highly polar, and in some reactions it behaves as germanium(II) chloride—

$$HGeCl_3 \rightleftharpoons GeCl_2 + HCl$$

When added to ether a bright-yellow insoluble oil is produced having the composition $(Et_2O)_2HGeCl_3$. This contrasts with the behaviour of trichlorosilane, which is unreactive; the difference may be ascribed to less π-bond character to the germanium—chlorine bonds making the hydrogen atom in trichlorogermane more protonic. The p.m.r. spectrum of the etherate shows, in addition to CH_3 and CH_2 resonances, a singlet at 14·6 p.p.m. downfield from tetramethylsilane, whereas in $HGeCl_3$ the proton resonance is only 7·6 p.p.m. downfield. This suggests that the proton is highly polarized and is indicative of an ionic structure such as (Nefedov and Manakov, 1966)—

$$(Et_2OH)^+(Et_2OGeCl_3)^-$$

Tribromogermane (m.p. −24°C) has been isolated by the action of hydrogen bromide on germanium at 450°C with the addition of copper powder; from aqueous hydrobromic acid and $Ge(OH)_2$, and by passing HBr through $HGeCl_3$. Distillation (30°C at 4 mm) of an ether solution of tribromogermane evidently gives a distillate containing $HGeBr_3$ and a yellow crystalline residue of germanium dibromide (Mironov and Gar, 1965a).

Tri-iodogermane has not been isolated, but if germanium di-iodide is dissolved in hydrogen chloride–ether, there is evidence for the formation of $(Et_2O)_2HGeClI_2$. Hydrogen iodide and trichlorogermane give germanium di-iodide. A great deal of use has been made of the addition of trichloro- and tribromo-germane to olefinic compounds; the course of these reactions depends strikingly on whether the ether complex is present.

3. *Germylsilane, H_3SiGeH_3*

This compound (b.p. 7°C; m.p. −119·7°C), which is stable at room temperature, was first obtained by passing a silent electric discharge through a mixture of germane and silane (Spanier and MacDiarmid, 1963). The most satisfactory preparative method is as follows—

$$Si_2H_6 \xrightarrow{KH} H_3SiK + SiH_4$$
$$H_3SiK + H_3GeCl \longrightarrow H_3SiGeH_3 + KCl$$

Its microwave spectrum (Table VI) has been examined by Varma and Cox (1967). The action of a silent electric discharge on mixtures of silane and germane or disilane and germane produces a wide variety of mixed hydrides, many of which can be partially identified from their vapour-phase chromatography retention times and mass spectra. Compounds reported include neo-$SiGe_4H_{12}$, iso-$Si_4Ge_2H_{14}$ and n-$SiGe_4H_{12}$ (Andrews and Phillips, 1966).

Some interesting chemistry has been carried out on related germanium–tin compounds (Wiberg *et al.*, 1967). Triphenylgermyltriphenyltin is converted into the corresponding acetoxy-compound in two stages by boiling with acetic acid—

$$Ph_3GeSnPh_3 \xrightarrow{HOAc} Ph_3GeSn(OAc)_3 \xrightarrow{HOAc} (AcO)_3GeSn(OAc)_3$$

A low-temperature reaction with hydrogen chloride followed by reduction with lithium aluminium hydride gives germylstannane as an extremely unstable ether solution—

$$(AcO)_3GeSn(OAc)_3 \xrightarrow[-50°C]{HCl} Cl_3GeSnCl_3 \xrightarrow{heat} GeCl_4 + SnCl_2$$

$$\downarrow LiAlH_4$$

$$H_3GeSnH_3 \xrightarrow{>-80°C} GeH_4 + Sn + H_2$$

4. *Germyl-amines, -phosphines and -oxides*

The compounds $(H_3Ge)_3N$, $(H_3Ge)_3P$, $(H_3Ge)_2O$ and H_3GeOMe have all been reported (Goldfarb and Sujishi, 1964; Cradock *et al.*, 1967a; Gibbon *et al.*, 1967; Cradock, 1968) and are of considerable theoretical interest in relation to carbon and silicon analogues.

Tertiary amines, phosphines and ethers all have roughly tetrahedral bond angles, whereas in trisilylamine it has been shown by electron diffraction (Hedberg, 1955) that the three silicon atoms and the nitrogen atom are essentially coplanar, and this structure is also supported by its i.r. and Raman spectra. If the lone-pair electrons on nitrogen donate into vacant *d* orbitals of silicon forming bonds of π-symmetry, then the geometry would be expected to change from pyramidal to planar. In $(H_3Si)_2O$ the large silicon—oxygen—silicon bond angle (144°) is also explicable in terms of d_π-p_π bonding (Almenningen *et al.*, 1963).

Analogous arguments might be expected to apply to compounds such as trisilyl-phosphine and -arsine, but electron-diffraction studies show convincingly that these are pyramidal, although vibrational spectra are compatible with planar structures (Beagley *et al.*, 1967). The evidence so far available on trisgermylamine and trisgermylphosphine suggests that they are non-planar, and in digermoxane the GeOGe angle is approximately tetrahedral.

Preparative methods developed for these compounds are as follows—

$$3H_3GeX + (H_3Si)_3P \longrightarrow (H_3Ge)_3P + 3H_3SiX$$

$$2H_3GeI + HgS \longrightarrow (H_3Ge)_2S + HgI_2$$

$$(H_3Ge)_2S + HgO \xrightarrow{-40°C} (H_3Ge)_2O + HgS$$

$$H_3GeCl + NaOMe \xrightarrow{-78°C} H_3GeOMe + NaCl$$

Methoxygermane is unstable even at low temperatures ($-23°C$) and decomposes to methanol and a yellow polymeric germanium hydride. It is hydrolysed to digermoxane—

$$2H_3GeOMe + H_2O \rightleftharpoons (H_3Ge)_2O + 2MeOH$$

Germylamine, germylphosphine and germylarsine have been described. The phosphine can be made by the reaction (Wingleth and Norman, 1967)—

$$LiAl(PH_2)_4 + 4H_3GeBr \xrightarrow[-45°C]{Et_2O} 4H_3GePH_2 + LiAlBr_4$$

If germane and phosphine or arsine are subjected to a silent electric discharge, low yields of germyl-phosphine or -arsine are obtained together with more complex compounds, such as Ge_2PH_7, Ge_3PH_9 and GeP_2H_6 (Drake and Jolly, 1962a). Trisgermylphosphine, $(H_3Ge)_3P$, is unstable; at room temperature it slowly decomposes to germane and a colourless liquid polymer of unknown structure. The germanium—phosphorus bond is rapidly cleaved by hydrogen halides. Germylphosphine and diborane combine at $-20°C$ (Drake and Riddle, 1968)—

$$2H_3GePH_2 + B_2H_6 \xrightarrow{-20°C} H_3GePH_2BH_3$$

Germylpotassium and diborane form a crystalline salt, potassium germyltrihydroborate, stable to $200°C$. It is fairly unreactive in aqueous alkaline solution, but decomposes in acid media (Rustad and Jolly, 1968)—

$$2H_3GeK + B_2H_6 \longrightarrow (H_3GeBH_3)K \xrightarrow{H^+} GeH_4 + H_2 + H_3BO_3$$

Germyl azide is stable below $-20°C$, its i.r. and Raman spectra strongly suggest that the molecular skeleton is non–linear—

$$Me_3SiN_3 + H_3GeF \rightarrow H_3GeN_3 + Me_3SiF$$

Digermylcarbodi-imide, which provides a convenient source of other germyl compounds, has been prepared by the reactions—

$$Me_3SiCl + AgCN_2 \rightarrow Me_3SiN{=}C{=}NSiMe_3$$
$$(Me_3SiN)_2C + H_3GeF \rightarrow (H_3GeN)_2C + 2Me_3SiF$$

With water it yields digermoxane, and with methanol, methoxygermane is formed. Its vibrational spectrum suggests that the GeNCNGe skeleton is non-linear (Cradock and Ebsworth, 1968).

The basicities of ethers and sulphides relative to silicon and germanium analogues have been invoked as providing evidence of d_π–p_π bonding. The technique is to compare changes in the oxygen–hydrogen stretching frequency of a weak acid in the presence of various electron donors. Phenol is often used, since at low concentrations (ca. 10^{-2} to 10^{-3} M) it gives a single sharp oxygen–hydrogen stretching frequency without the broad hydrogen-bonded ν(O–H) to lower frequency. Addition of a base, such as dimethyl ether, produces a broad hydrogen-bonded O—H stretch due to an equilibrium concentration of the hydrogen-bonded adduct, the difference in frequency being proportional to the strength of the base—

$$PhOH + Me_2O \rightleftharpoons PhO—H \cdots OMe_2$$

Since carbon is more electronegative than silicon or germanium, silyl- and germyl-oxides and -sulphides should be stronger bases than ethers or organic sulphides in the absence of π-bonding between d orbitals of silicon or germanium and doubly occupied p-orbitals of oxygen or sulphur. The experimental observations show that $(H_3Si)_2O$ and H_3SiOMe are less basic than Me_2O, and hence the silicon—oxygen bond has significant d_π–p_π character. Digermoxane and methoxygermane by contrast, are more basic than dimethyl ether, which clearly implies that d_π–p_π bonding between germanium and oxygen is less than for silicon—oxygen analogues.

Similar studies on the sulphides $(H_3Si)_2S$ and H_3SiSMe reveal that these and the germyl analogues $[(H_3Ge)_2S$ and $H_3GeSMe]$ are much weaker bases than Me_2S, showing negligible hydrogen bonding to phenol; the inference being that both silicon—sulphur and germanium—sulphur bonds involve significant π-bonding (Wang and Van Dyke, 1968).

J. Inorganic germanium–nitrogen compounds

1. *Reactions involving ammonia and amines*

Germanium(II) nitride, amine co-ordination complexes and a phthalo-cyanine derivative have already been discussed. The reactions between germanium halides and ammonia and various primary and secondary amines have received some study, but little is known about the structure and properties of the products. Secondary amines are simplest in their reactions with germanium tetrachloride—

$$GeCl_4 + 8R_2NH \rightarrow Ge(NR_2)_4 + 4R_2NH_2Cl$$

Anderson (1952a) prepared a series of tetrakisdialkylaminogermanes, most of which were colourless mobile liquids, very readily hydrolysed. $Ge(NEt_2)_4$ decomposes at its boiling point, giving a deep red liquid. Similarly NN'-dialkylethylenediamines react with germanium tetra-chloride in boiling benzene to give spiro-imidazolidines with germanium as the central atom—

$$
\begin{array}{ccc}
R & & R \\
| & & | \\
{-}N & & N{-} \\
& \diagdown \diagup & \\
& Ge & \\
& \diagup \diagdown & \\
{-}N & & N{-} \\
| & & | \\
R & & R
\end{array}
$$

These compounds are more sensitive to hydrolysis than the silicon analogues (Yoder and Zuckerman, 1964). Their p.m.r. spectra show a single ring-methylene resonance, indicating either planarity of the ring system or, more probably, rapid inversion of tetrahedral nitrogen. A pyrrole derivative has been obtained as a yellow solid (m.p. 202°C) by the reaction (Schwarz and Reinhardt, 1932)—

$$GeCl_4 + 4C_4H_4NK \longrightarrow \left[Ge{-}N\diagup\Box \right]_4 + 4KCl$$

The reactions of primary amines (Thomas and Southwood, 1931) and ammonia with germanium tetrahalides have also been studied (Thomas and Pugh, 1931; Johnson and Sidwell, 1933; Dennis and Work, 1933)—

$$GeX_4 + RNH_2 \rightarrow [Ge(NHR)_2]_n + RNH_3X$$
$$GeX_4 + NH_3 \rightarrow [Ge(NH)_2]_n + NH_4X$$

The imide, which is clearly polymeric, is described as an involatile white powder rapidly decomposed by water, but no detailed studies have been reported. Chlorogermanes behave differently, owing to the ease with which ammonia can be eliminated from intermediates—

$$H_3GeCl + NH_3 \xrightarrow{-50°C} NH_4Cl + GeH_4 + (GeH)_n$$
$$H_2GeCl_2 + 2NH_3 \xrightarrow{-78° \text{ to } -27°C} 2NH_4Cl + Ge$$

Germanium tetracyanide is discussed under pseudohalides (Chapter 7).

2. Germanium(IV) nitride

This compound, Ge_3N_4, has been prepared by two reactions (Johnson, 1930; Schwarz and Schenk, 1930)—

$$Ge + NH_3 \xrightarrow{700°C} Ge_3N_4 \xleftarrow{400°C} [Ge(NH)_2]_n$$

When pure, it forms a white powder that is extremely inert. It does not react with hot mineral acids or hot caustic soda solution, but may be reduced or chlorinated at high temperatures. At 1000°C, it decomposes to germanium and nitrogen—

$$Ge_3N_4 \diagup\begin{array}{l} \xrightarrow[700°C]{H_2} 3Ge + 4NH_3 \\ \\ \xrightarrow[600°-700°C]{Cl_2} GeCl_4 + 2N_2 \end{array}$$

3

3 | TETRAORGANOGERMANES

The methods of forming germanium–carbon bonds are described together with a discussion of those reactions that do not feature in subsequent Chapters. The high thermal stability and low chemical reactivity of organogermanes means that a wide variety of preparative methods are applicable. Although there is some evidence for germonium ions, solvolysis of organogermanes and organogermanium halides are reactions of the S_N2 type.

I. SYMMETRICAL COMPOUNDS

Over 40 symmetrical organomonogermanes have been described, and with few exceptions they have always been prepared by the alkyl- or aryl-ation of germanium tetrahalides—usually the tetrachloride. Several indirect methods may also be used [e.g., $(Me_3Ge)_2O + MeLi \rightarrow Me_4Ge + Me_3GeOLi$; $Et_3GeH + C_2H_4 \rightarrow Et_4Ge$; $Et_3GeC\vdots CH + 2H_2 \rightarrow Et_4Ge$; $Ph_3GeLi + PhBr \rightarrow Ph_4Ge$], but these are not of preparative importance.

Grignard reagents provide the most versatile means of establishing metal–carbon bonds, and the reaction was first applied to tetraorganogermanes in 1925 by Tabern et al. and by Morgan and Drew—

$$GeCl_4 \underset{\xrightarrow{EtMgBr}\ Et_4Ge}{\overset{\xrightarrow{PhMgBr}\ Ph_4Ge}{}}$$

The reactions are very simple experimentally; a large excess of the Grignard reagent is usually employed and the reactants are boiled under reflux for periods of up to several days, usually in ether or ether–benzene solution. Since organogermanes are stable to air and water, isolation of the product presents no problem. The choice of solvent is occasionally dictated by the boiling point of the tetraorganogermane. For example, tetramethylgermane boils at 43·4°C and separation from diethyl ether is difficult, so it is best prepared in dibutyl ether solution, from which it can be isolated in almost quantitative yield by direct distillation without hydrolysis (Gladshtein et al., 1959). Tetrahydrofuran has been used in cases such as the preparation of (vinyl)$_4$Ge, where the formation of the Grignard reagent does not proceed readily in diethyl ether (Seyferth, 1957).

The reaction of germanium tetrachloride with Grignard reagents is a stepwise process and the rate of the different stages can vary greatly—

$$GeCl_4 \xrightarrow{RMgX} RGeCl_3 \xrightarrow{RMgX} R_2GeCl_2 \xrightarrow{RMgX} R_3GeCl \xrightarrow{RMgX} R_4Ge$$

Although relative rates are not known, there is a considerable body of evidence that the first two or three R groups are introduced more readily than the fourth. With aryl Grignard reagents it is common practice to replace part of the ether by a higher-boiling solvent, such as benzene or toluene, in order to increase the reaction temperature and hence the rate of the last stage.

The yields obtained in the Grignard synthesis vary widely; simple tetra-alkyls and -aryls are formed in 70–100% yield, but almost invariably side reactions occur and, especially when organic groups having a large steric requirement are involved, the yield of tetraorganogermane can be extremely small. The reason for low yields, apart from experimental difficulties in isolation or purification, stems primarily from slow substitution of the halogen in R_3GeX, which either remains up to hydrolysis or takes part in other reactions. Most of these complications are discussed in other Chapters, but the two types of side reaction are the formation of oxides and hydroxides after hydrolysis owing to incomplete reaction and the production of hydrides, digermanes and polygermanes due to the formation of germyl Grignard reagents or to reductive coupling processes. Even in simple cases like ethylmagnesium bromide, the digermane Et_6Ge_2 is formed in isolable amount (Gilman *et al.*, 1959). Tetraphenylgermane is formed in much higher yield if free magnesium is excluded from the reaction mixture, thereby minimizing the formation of the digermane (Glockling and Hooton, 1966). There are several examples where even prolonged periods of reflux fail to substitute the fourth halogen (Bauer and Burschkies, 1932; West, 1952; Glockling and Hooton, 1962b). Some support for the suggestion that steric factors influence the substitution of the fourth halogen atom comes from the effect of other Grignard reagents on tricyclohexylbromogermane. Methyl-, ethyl- and benzyl-Grignard reagents react, whereas the more sterically hindered isopropyl- and cyclohexyl-Grignard reagents do not—

$$3C_6H_{11}MgBr + GeCl_4 \longrightarrow (C_6H_{11})_3GeBr \xrightarrow{MeMgI} (C_6H_{11})_3GeMe$$

Some caution is needed in relating reactivity to steric factors; thus phenyl-magnesium bromide is apparently unreactive towards tricyclohexylbromo-germane, whereas it reacts readily with tertiarybutyltrichlorogermane—

$$Bu^tGeCl_3 \xrightarrow{PhMgBr} Bu^tGePh_3$$

The reaction of isopropylmagnesium chloride with germanium tetra-chloride is of interest, since some 25 volatile products have been detected by vapour-phase chromatography, and only 5–20% is accountable as tetraiso-propylgermane (Carrick and Glockling, 1966). A model of this molecule suggests that four isopropyl groups can pack around the central germanium without gross steric interference. In this reaction, complications, additional to those already discussed, include isomerization of isopropyl- to n-propyl groups at some stage in the reaction leading to tri-isopropyl-n-propyl-germane.

Organolithium reagents have been less used in the synthesis of sym-metrical tetraorganogermanes, and in most cases much lower yields are obtained (e.g., Et_4Ge, 12%; $[Ph(CH_2)_3]_4Ge$, 33%; $(n-C_{10}H_{21})_4Ge$, 18%; $(m-FC_6H_4)_4Ge$, 25%; (1-naphthyl)$_4Ge$, 38%) (Fuchs et al., 1956; Gilman et al., 1959, 1962b). The reasons for low yields are not clear; they may be due in part to difficulties in handling organolithium reagents so that oxida-tion and hydrolysis take place. There seems to be no reason why metal–halogen exchange reactions should not occur and, particularly if this happens at an early stage (e.g., giving an intermediate such as Cl_3GeLi), the possi-bilities for side reactions are considerable.

By products may also result from reductive processes giving ger-manium(II) intermediates. It is significant that trialkyl- and triaryl-halo-germanes that cannot be reduced to germanium(II) compounds are alkylated in high yield by lithium alkyls. Another complication is due to the cleavage of ethereal solvents by lithium alkyls. Phenyl-lithium in toluene gives a high yield (90%) of tetraphenylgermane (Johnson and Nebergall, 1949).

Organoaluminium compounds, several of which (Me_3Al, Et_3Al, Bu^i_3Al) have become commercially available because of their application to olefin dimerization and telomerization reactions as well as the formation of iso-tactic polymers from olefins, may be used quite effectively in the synthesis of symmetrical tetraorganogermanes, but the ease with which they take fire in air makes them treacherous to handle (Zakharkin and Okhlobystin, 1961; Glockling and Light, 1967).

One respect in which these reactions contrast with the corresponding Grignard reactions is that the slow stage is the monoalkylation of ger-manium tetrachloride so that, commonly, the only monogermanium pro-ducts isolated are the tetra-alkyl and unreacted tetrachloride. This is surprising since the initial reaction is quite exothermic and it suggests that germanium tetrachloride is deactivated possibly by formation of penta- or hexa-co-ordinate halogen-bridged species, such as $(R_2AlCl)GeCl_4$ or $(R_2AlCl)_2GeCl_4$. Since aluminium-alkyls and -alkyl halides are usually dimeric, many equilibria must be involved in these alkylation reactions, e.g.—

$$2GeCl_4 + (R_3Al)_2 \rightleftharpoons 2RGeCl_3 + (R_2AlCl)_2$$
$$2R_3GeCl + (RAlCl_2)_2 \rightleftharpoons 2R_4Ge + Al_2Cl_6$$

It is clear that the equilibria favour the formation of tetra-alkylgermanes since high yields are obtained even when the aluminium alkyl is present in only small excess. Yields are commonly in the range 70–90%.

The reactions are best carried out without solvent, and the system Me_3Al–$GeCl_4$–$NaCl$ is the most satisfactory in that tetramethylgermane, the most volatile component, may be distilled directly from the mixture. Methylaluminium sesquichloride alone does not alkylate germanium tetrachloride appreciably, but with sodium chloride present it behaves like trimethylaluminium. The effect of adding sodium chloride is to increase the rate of formation of the tetra-alkylgermane without increasing the yield appreciably—

$$3GeCl_4 + 4R_3Al + 4NaCl \rightarrow 3R_4Ge + 4NaAlCl_4$$

Side reactions occur in much the same way as has been discussed under Grignard reagents, giving di- and poly-germanes. These are considered in greater detail under catenated organogermanes (Chapter 8). With tri-isobutylaluminium the proportion of isobutylpolygermanes is high, and if the reaction is carried out at high temperature (140°–150°C), various iso-butylgermanium hydrides are formed by the reactions—

$$2Bu^i_3Al \rightarrow (Bu^i_2AlH)_2 + 2C_4H_8$$
$$(Bu^i_2AlH)_2 + Bu^i_nGeCl_{4-n} \rightarrow Bu^i_nGeH_{4-n}$$

The rate of reduction of germanium—chlorine bonds by the alkyl aluminium hydride is so much greater than the rate of alkylation that some monogermane is produced.

The Wurtz–Fittig reaction has been applied to the preparation of tetra-arylgermanes (Schwarz and Lewinsohn, 1931; Worral, 1940)—

$$4PhBr + GeCl_4 + 8Na \xrightarrow{Et_2O} Ph_4Ge + 4NaCl + 4NaBr$$

Zinc dialkyls have also been used, but the inconvenience of having to prepare these spontaneously inflammable starting materials makes it an unattractive method with no advantages over those already discussed. Both tetramethyl- and tetraethyl-germane have been made this way (Dennis and Hance, 1925; Lengel and Dibeler, 1952). Of greater interest is the use of zinc diaryls formed *in situ* by the addition of zinc chloride to Grignard reagents followed by the addition of germanium tetrachloride (Simons *et al.*, 1933). As applied to the tolyl Grignard reagents, this procedure evidently gives the tetra-tolylgermanes even for the *o*-tolyl, whereas the Grignard reagent alone gives (*o*-tolyl)$_3$GeBr. If this is due to the formation of an equilibrium concentration of (*o*-tolyl)$_2$Zn or (*o*-tolyl)ZnBr, then it is surprising, since both types of organozinc compounds are generally less reactive than Grignard reagents in nucleophilic substitution reactions.

II. UNSYMMETRICAL COMPOUNDS

All types of unsymmetrical germanes (A_3BGe, A_2B_2Ge, A_2BCGe and $ABCDGe$) are known, and the methods of preparation cover almost all aspects of organogermanium chemistry. The most important method is the alkylation or arylation of organogermanium halides by organometallic reagents. In many cases, the greater reactivity of lithium alkyls over Grignard reagents enables sterically hindered compounds to be prepared. For example tricyclohexylbromogermane, which is unreactive towards phenylmagnesium bromide, is phenylated by phenyl-lithium (Johnson and Nebergall, 1949)—

$$(C_6H_{11})_3GeBr + PhLi \rightarrow (C_6H_{11})_3GePh + LiBr$$

The importance of steric factors in this reaction is illustrated by the failure of isopropyl-lithium to substitute the remaining bromine atom. The reaction of mixtures of Grignard reagents and germanium tetrachloride has been studied by means of gas-chromatographic techniques and is shown to produce tetra-alkylgermanes in roughly statistical proportions (Semlyen *et al.*, 1964). Functionally substituted mercury alkyls are sufficiently reactive to alkylate germanium halides and hydrides (Baukov and Lutsenko, 1964; Baukov *et al.*, 1965a, b, 1966)—

$$2Bu^n_3GeI + Hg(CH_2CO_2Pr^n)_2 \rightarrow 2Bu^n_3GeCH_2CO_2Pr^n + HgI_2$$
$$2Et_3GeI + Hg(CH_2COMe)_2 \rightarrow 2Et_3GeCH_2COMe + HgI_2$$

Mercuric iodide can be separated as the insoluble pyridine complex. Similar reactions with germanium hydrides produce mercury—

$$Bu^n_2GeH_2 + 2Hg(CH_2CO_2Me)_2 \rightarrow Bu^n_2Ge(CH_2CO_2Me)_2 + 2Hg + 2CH_3CO_2Me$$

These reactions proceed smoothly and in high yield, as do exchange reactions between substituted tin alkyls and germanium halides—

$$R_3SnCH_2COMe + R_3GeX \rightarrow R_3GeCH_2COMe + R_3SnX$$

Wurtz reactions have been applied to the synthesis of unsymmetrical germanes, usually by adding a mixture of the halides to a dispersion of the alkali metal in a boiling hydrocarbon (Eaborn and Pande, 1960b), but in all cases some digermane is also formed—

$$Et_3GeBr + PhCl \xrightarrow{\text{Na}} Et_3GePh$$

In some syntheses, the pre-formed sodium alkyl has been used (Bott *et al.*, 1963a)—

$$Me_3GeBr + Ph_3CNa \rightarrow Me_3GeCPh_3 + NaBr$$

The addition of germanium hydrides to alkenes and alkynes is a versatile route to unsymmetrical organogermanes (Chapter 4). Many other derivatives can be converted into unsymmetrical germanes; these include the more reactive germanium–metal compounds discussed in Chapter 8. For example, triphenylgermylsodium and alkyl- or aryl-halides combine readily (Smith and Kraus, 1952; Tamborski *et al.*, 1962)—

$$Ph_3GeNa \begin{cases} \xrightarrow{PhCH_2Cl} Ph_3GeCH_2Ph \\ \xrightarrow{Br(CH_2)_5Br} Ph_3Ge(CH_2)_5GePh_3 \end{cases}$$

Triorganogermyl–alkali metal compounds add to terminal and non-terminal alkenes, though less readily than triphenylsilyl-lithium (Gilman and Gerow, 1957b; Vyazankin *et al.*, 1966b)—

$$Ph_3GeLi + Ph_2C:CH_2 \longrightarrow Ph_3GeCH_2C(Li)Ph_2 \xrightarrow{H_2O} Ph_3GeCH_2CHPh_2$$

$$Et_3GeLi + RCH:CH_2 \longrightarrow Et_3GeCH_2CH(Li)R \xrightarrow{H_2O} Et_3Ge(CH_2)_2R$$

Some unexpected variations in reactivity have been reported; thus *trans*-stilbene is unreactive towards triphenylgermyl-lithium, whereas it reacts with the potassium salt—

$$PhCH:CHPh \xrightarrow[\text{2. } H_2O]{\text{1. } Ph_3GeK} Ph_3GeCH(Ph)CH_2Ph$$

The addition of triphenylgermyl-lithium to benzophenone differs from that of the silyl analogue—

$$Ph_2C:O \begin{cases} \xrightarrow[\text{2. } H_2O]{\text{1. } Ph_3GeLi} Ph_3GeC(OH)Ph_2 \\ \xrightarrow[\text{2. } H_2O]{\text{1. } Ph_3SiLi} Ph_3SiOCHPh_2 \end{cases}$$

The reason for this difference is not known; it is possible that the same initial product is formed $(Ph_3MC(OLi)Ph_2)$ and that the silicon compound subsequently rearranges, thereby forming the much stronger silicon–oxygen bond. Triphenylgermyl-lithium also adds normally to formaldehyde (Gilman and Gerow, 1955c).

Asymmetric organogermanes

The study of substitution reactions on asymmetric germanes is of great interest in relation to reaction mechanisms, and the available information reveals close similarities to the reactions of asymmetric organosilanes (Sommer, 1962). Two partial resolutions of asymmetric germanes, and several substitution reactions have been reported. (Eaborn *et al.*, 1966a; Eaborn and Varma, 1967). The synthetic routes illustrate various methods of forming and cleaving germanium–carbon bonds—

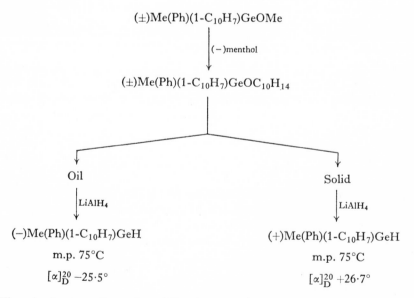

$$Ph_3GeEt \xrightarrow{Br_2} Ph_2(Et)GeBr \xrightarrow{Pr^iM_gBr} Ph_2(Et)GePr^i$$

A preliminary account of a more extensive study of substitution reactions on asymmetric organogermanes has been given by Brook and Peddle (1963). Resolution of racemic methyl-1-naphthylphenylmethoxygermane by conversion to the (−)menthol derivative has been achieved according to the following scheme—

The absolute configuration (R) of the corresponding silane (+)Me(Ph)-(1-C$_{10}$H$_7$)SiH has been unambiguously determined; it forms a solid solution with the dextrorotatory germane melting over less than a 1°C range, whereas a mixture of (+)GeH and (−)SiH melts over a wide range. Hence the dextrorotatory germane has the R configuration. All of the substitution reactions reported, with the exception of lithium aluminium hydride reduction of the chloride proceed with retention of configuration—

$$(-)\text{GeH} \xrightarrow[\text{retention}]{\text{Cl}_2} (+)\text{GeCl} \xrightarrow[\text{inversion}]{\text{LiAlH}_4} (+)\text{GeH}$$

These correlations are based on the mixed-melting behaviour with the corresponding silanes of known configuration, and on the application of Brewster's rules of atomic asymmetry. Two noteworthy features are the high stability of the lithiogermane, which implies close ion-pair association, and that lithium aluminium hydride reduction of the methoxygermane proceeds with retention of configuration—

$$(-)\text{GeH} \xrightarrow{\text{BuLi}} \text{GeLi} \xrightarrow{\text{CO}_2} (+)\text{GeCO}_2\text{H}$$

$$\text{H}_2\text{O} \downarrow \qquad\qquad \downarrow \text{Ph}_2\text{CO}$$

$$(-)\text{GeH} \qquad\qquad (+)\text{GeC(OH)Ph}_2$$

$$(+)\text{GeCO}_2\text{H} \xrightarrow{\text{CH}_2\text{N}_2} (+)\text{GeCO}_2\text{Me} \xrightarrow{\text{heat}} (-)\text{GeOMe} \xrightarrow{\text{LiAlH}_4} (-)\text{GeH}$$

The asymmetric germane, $1\text{-C}_{10}\text{H}_7(\text{Me})\text{PhGeCl}$ is optically stable in hydrocarbon and chlorinated solvents, but in strongly basic ethereal media (such as tetrahydrofuran) racemization is rapid. The mechanism probably involves co-ordination of one or two molecules of solvent around germanium (Carre *et al.*, 1968).

III. ALKENYLGERMANES

Vinyl- and substituted vinyl-germanes can be obtained by the traditional Grignard or organolithium methods (Seyferth and Vaughn, 1963)

$$\text{Ph}_3\text{GeCH:CH}_2 \xleftarrow{\text{Ph}_3\text{GeBr}} \text{CH}_2\text{:CHLi} \xrightarrow{\text{GeCl}_4} \text{Ge(CH:CH}_2)_4$$

Application of this method to propenyl-lithium produces *cis* and *trans* isomers, which have been separated and characterized spectroscopically—

$$\text{Me}_3\text{GeBr} + \text{MeCH:CH}_2\text{Li} \longrightarrow \underset{H}{\overset{\text{Me}_3\text{Ge}}{>}}\text{C=C}\underset{\text{Me}}{\overset{H}{<}} \;+\; \underset{H}{\overset{\text{Me}_3\text{Ge}}{>}}\text{C=C}\underset{H}{\overset{\text{Me}}{<}}$$

3*

Equilibration of these isomers in tetrahydrofuran is catalysed by lithium giving 92 % of the *trans* and 8 % of the *cis* form. The mechanism of the inter-conversion is considered to involve a radical–anion intermediate. The electron transferred from lithium will occupy a π-antibonding orbital of the olefin and weaken the carbon—carbon π-bond, thus reducing the energy barrier to rotation about this bond—

$$\text{Me}_3\text{GeCH:CHMe} \underset{cis}{\overset{\text{Li}}{\rightleftharpoons}} \left[\begin{array}{c} \text{Me}_3\text{Ge}\diagdown \qquad \diagup\text{Me} \\ \qquad \text{C=C} \\ \text{H}\diagup \qquad \diagdown\text{H} \end{array} \right]^{-\bullet} \text{Li}^+ \overset{\text{Li}}{\rightleftharpoons} \underset{trans}{\text{Me}_3\text{GeCH:CHMe}}$$

Alkali-metal-induced isomerization of olefins is rare, and for germyl- and silyl-alkenyls the radical anion may be stabilized by π-bonding to the metal. Solvation effects must also be important, since in ethereal solution isomeri-zation does not occur (Seyferth *et al.*, 1966). Both *cis*- and *trans*-tetra-propenylgermanes have been isolated by the reaction (Nesmeyanov *et al.*, 1965b)—

$$cis\text{- or } trans\text{-MeCH:CHLi} \xrightarrow{\text{GeCl}_4} cis\text{- or } trans\text{-(MeCH:CH)}_4\text{Ge}$$

The Wurtz reaction has been applied to the synthesis of tetra-β-styryl-germane, though in low yield (Birr and Kraeft, 1961)—

$$\text{GeCl}_4 + \text{PhCH:CHBr} \xrightarrow{\text{Na}} \text{Ge(CH:CHPh)}_4$$

The addition of germanium hydrides to alkynes is discussed in Chapter 4, but metal hydride reduction of alkynylgermanes is a stereospecific *trans* addition—

$$\text{Et}_3\text{GeC:CPh} \xrightarrow[\text{2. } \text{H}_2\text{O}]{\text{1. Bu}^n_2\text{AlH}} \begin{array}{c} \text{Et}_3\text{Ge}\diagdown \qquad \diagup\text{H} \\ \qquad \text{C=C} \\ \text{H}\diagup \qquad \diagdown\text{Ph} \end{array}$$

The suggested mechanism is the formation of a π-complex between alu-minium and the alkyne followed by attack by R_2AlH from the opposite side of the complex (Eisch and Foxton, 1968)—

$$\begin{array}{c} \text{R}_3\text{GeC}\equiv\text{CPh} \\ \downarrow \\ \text{R}_2\text{AlH} \end{array} \longrightarrow \begin{array}{c} \text{R}_3\text{Ge}\diagdown \qquad \diagup\text{HAlR}_2 \\ \qquad \text{C=C} \\ \text{R}_2\text{AlH}\diagup \qquad \diagdown\text{Ph} \end{array}$$

β-Haloethylgermanes have been dehydrohalogenated to the alkenyls, although this reaction is usually in competition with β-elimination (Mazerolles, 1960)—

$$\text{Bu}^n_3\text{GeCHBrCH}_2\text{Br} \xrightarrow[100°\text{C}]{\text{Et}_2\text{NH}} \text{Bu}^n_3\text{GeC(Br):CH}_2$$

In much the same way, germylcarbinols dehydrate over such reagents as potassium hydrogen sulphate or phosphorus halides (Gverdtsiteli et al., 1964)—

$$Ph_3GeC(OH)(Me)Ph \xrightarrow{PBr_3} Ph_3GeC(Ph):CH_2$$

The low reactivity of vinylgermanes is shown by some of the reactions that proceed without destroying the GeC:C grouping (Mironov et al., 1966)—

$$Me_3GeCH:CHCl + Me_3GeCl \xrightarrow{Na} Me_3GeCH:CHGeMe_3$$

$$Me_3GeCCl:CH_2 + Me_3GeCl \xrightarrow{Na} (Me_3Ge)_2C:CH_2$$

Oxidation of alkenylgermanes produces the corresponding carboxylates or ketones in low yield.

Various reagents have been added to the carbon—carbon double bond. Vinylgermanes are more reactive towards addition of dichlorocarbene than vinylsilanes (Seyferth and Dertouzos, 1968), whereas methylene iodide and diazoacetic ester both form cyclopropylgermanes (Seyferth and Cohen, 1962; Dolgii et al., 1963)—

$$Me_3GeCH:CH_2 + CH_2I_2 \xrightarrow{Zn/Cu} Me_3Ge\!-\!\!\triangleleft$$

$$Et_3GeCH:CH_2 \xrightarrow{PhHgCCl_2Br} Et_3Ge\!-\!\!\triangledown\!\!\cdot Cl_2$$

Triphenylvinylgermane adds phenyl-lithium to give, after hydrolysis, the β-phenylethyl compound. By contrast triphenylvinyl-lead is cleaved by phenyl-lithium, yielding tetraphenyl-lead and vinyl-lithium (Seyferth and Weiner, 1962)—

$$Ph_3GeCH:CH_2 \xrightarrow{PhLi} Ph_3GeCH(Li)CH_2Ph \xrightarrow{H_2O} Ph_3GeCH_2CH_2Ph$$

Bromine and thioglycollic acid add to vinylgermanes, and several free-radical-induced addition reactions have been reported (Mazerolles and Lesbre, 1959)—

$$Et_3GeCH:CH_2 \xrightarrow{HSCH_2CO_2H} Et_3Ge(CH_2)_2SCH_2CO_2H$$

$$Et_3GeCH:CH_2 \xrightarrow[(PhCO)_2O_2]{CCl_4} Et_3GeCHClCH_2CCl_3$$

$$Et_3GeCH:CH_2 \xrightarrow[(PhCO)_2O_2]{BrCH_2CO_2Et} Et_3GeCH(Br)(CH_2)_2CO_2Et$$

Cyclopentadienyltriethylgermane can be reduced catalytically without affecting the germanium—carbon bonds, and it also undergoes a normal Diels–Alder reaction with maleic anhydride (Lesbre *et al.*, 1962).

β-Alkenylgermanes, $R_3GeCH_2CH\!:\!CHR'$, show a high reactivity in ionic addition reactions and an unusually high intensity of Raman lines associated with the stretching vibration of the double bond. These anomalies and the bathochromic shift in their u.v. spectra have been tentatively explained in terms of an electronic interaction (other than an inductive effect) between the metal and the π-component of the double bond (Leites *et al.*, 1964).

IV. ALKYNYLGERMANES

All of the group IV elements form alkynyl–metal compounds of the type $R_nM(C\!:\!CR')_{4-n}$; their chemistry has been reviewed by Davidsohn and Henry (1967). Germanium forms a wide range of alkynyls, they are more readily formed than alkylgermanes and are rather more readily solvolysed. Few alkynyls present any explosive hazard, though tetraethynylgermane explodes on rapid heating and is sensitive to friction.

Preparative methods are straightforward, and either alkali metal ethynyls or Grignard derivatives may be used (Eaborn and Walton, 1964; Ibekwe and Newlands, 1965; Davidsohn and Henry, 1966; Sladkov and Luneva, 1966)—

$$GeCl_4 + HC\!:\!CNa \xrightarrow{\text{THF}} Ge(C\!:\!CH)_4$$

$$R_nGeCl_{4-n} + R'C\!:\!CNa \xrightarrow{\text{THF}} R_nGe(C\!:\!CR')_{4-n}$$

$$(R' = H, \text{alkyl or aryl})$$

Some complications are observed, probably owing to reaction with di-sodioacetylene—

$$Ph_3GeCl + Na_2C_2 \rightarrow Ph_3GeC\!:\!CGePh_3 \; (55\%)$$

Reactions of this type have been elegantly applied by Steingross and Zeil, (1966); these illustrate the high stability of the germanium—carbon bond in ethynylgermanes—

$$\textit{trans-}CHCl\!:\!CHCl \xrightarrow{\text{MeLi}} LiC\!:\!CCl \xrightarrow{\text{Me}_3\text{GeX}} Me_3GeC\!:\!CCl \xrightarrow{\text{PhLi}}$$

$$Me_3GeC\!:\!CLi \xrightarrow{\text{H}_2\text{O}} Me_3GeC\!:\!CH$$

Further variations have been used to produce "mixed" metal ethynyls (Findeiss *et al.*, 1967)—

$$Ph_3SiC\!:\!CLi \xrightarrow{\text{Ph}_3\text{GeBr}} Ph_3SiC\!:\!CGePh_3$$

The high reactivity of tin–nitrogen compounds towards protic reagents may also be applied—

$$Me_3SnNEt_2 + HC\vdots CNa + Me_3GeCl \xrightarrow{THF} Me_3SnC\vdots CGeMe_3$$

Ethynyltin compounds are less stable and more reactive than the corresponding germanes, and exchange reactions can be carried out under fairly mild conditions (Sharanina et al., 1966)—

$$Et_3SnC\vdots CR + GeCl_4 \xrightarrow{150°-200°C} RC\vdots CGeCl_3 + R_3SnCl$$

A further illustration of the stability of these germanes is the dehydrobromination of alkenes by sodamide (Mazerolles, 1960)—

$$Bu^n_3GeC(Br)\vdots CH_2 \xrightarrow[150°C]{NaNH_2} Bu^n_3GeC\vdots CH$$

Many ethynylgermanes are sufficiently stable to be sublimed or vacuum distilled. They are unaffected by aqueous acids, alkalis or thiols. Mild bromination to the alkenyls can be achieved without cleavage. Reaction of diphenyldiethynylgermane with diazomethane yields the dipyrazoyl derivative—

$$Ge(C\vdots CH)_4 \xrightarrow{Br_2} Ge(CBr\vdots CHBr)_4$$

$$Ph_2Ge(C\vdots CH)_2 \xrightarrow{CH_2N_2} Ph_2Ge\left[\underset{\underset{H}{N}}{\begin{array}{c} \\ \end{array}}\right]_2$$

Some compounds have been reported in which the ethynyl group is remote from the germanium—carbon bond; these behave like organic ethynyls, but propargylgermanes are anomalous in being cleaved by carboxylic acids (Mazerolles, 1960)—

$$Bu^n_3GeCH_2C\vdots CH + RCO_2H \rightarrow Bu^n_3GeOCOR + MeC\vdots CH$$

In the formation of propargylgermanes from propargyl chloride, magnesium and the germanium halide, the isomeric allenylgermane is also produced in about 20% yield (Masson et al., 1967)—

$$R_3GeX + CH\vdots CCH_2Br \xrightarrow{Mg} R_3GeCH_2C\vdots CH + R_3GeCH\vdots C\vdots CH_2$$

V. HETEROCYCLIC ORGANOGERMANES

Grignard reagents derived from 1,4-dibromobutane or 1,5-dibromopentane react with germanium tetrachloride or dihalides of the type

R_2GeX_2, giving products in which germanium forms part of a 5- or 6-membered ring (Schwartz and Reinhardt, 1932; Mazerolles, 1962; Mazerolles et al., 1966)—

$$Et_2GeBr_2 + (CH_2)_4(MgBr)_2 \longrightarrow Et_2Ge\langle\text{ring}\rangle$$

$$GeCl_4 + (CH_2)_5(MgBr)_2 \longrightarrow \langle\rangle GeCl_2 + \langle\rangle Ge \langle\rangle$$

In the second example the spiro-germane may be separated from the dichlorogermanacyclohexane by treating the mixture with lithium aluminium hydride, thereby forming the more volatile dihydride, cyclo-$(CH_2)_5GeH_2$, without affecting the spiran. This method of synthesis is only successful for 5- and 6-membered rings. Seven and higher membered rings have not yet been reported, but the germanacyclobutane ring system has been obtained by a Wurtz reaction—

$$Bu^n_2GeCl_2 + (CH_2)_3Cl_2 \xrightarrow[\text{xylene}]{Na} Bu^n_2Ge\langle\rangle \qquad 10\%$$

$$Bu^n_2Ge\langle\overset{}{\underset{Cl}{}}CH_2Cl \xrightarrow[\text{xylene}]{Na} Bu^n_2Ge\langle\rangle \qquad 75\%$$

A further preparative method that has not been extensively explored is the reaction between a divinylgermane and an organogermanium dihydride—

$$R_2Ge(CH:CH_2)_2 + R'_2GeH_2 \longrightarrow R_2Ge\langle\rangle GeR'_2$$

These compounds illustrate the effect of ring size and type of group on chemical reactivity. Thus bromine will cleave a germanium—phenyl bond in preference to a 5-membered ring, and a 4-membered ring is even more reactive—

$$\square GePh_2 \xrightarrow[0°C]{Br_2,\ EtBr} \square Ge(Br)Ph \xrightarrow{Br_2,\ EtBr} \square GeBr_2$$

$$\square Ge\langle\rangle + Br_2 \xrightarrow[0°C]{EtBr} \square Ge\overset{(CH_2)_3Br}{\underset{Br}{}}$$

Halogen acids and sulphuric acid also cleave the heterocyclic rings, and the germanacyclobutane ring is cleaved by hot alcoholic potassium hydroxide—

R_2Ge (germanacyclobutane) $\xrightarrow{HX} R_2Ge(X)Pr^n$

$\xrightarrow{KOH} (R_2GePr^n)_2O$

R_2Ge (ring) $\xrightarrow{H_2SO_4} R_2Bu^nGeOSO_3H \xrightarrow{NaOH} (R_2Bu^nGe)_2O$

A further illustration of the greater reactivity of the germanacyclobutane ring is its cleavage by lithium aluminium hydride, trialkylsilanes and alkylchlorosilanes. The silane reactions require free-radical initiation (Mazerolles and Dubac, 1967)—

R_2Ge $\xrightarrow{LiAlH_4} R_2(Pr^n)GeH$

$\xrightarrow{Et_3SiH} R_2Ge(H)(CH_2)_3SiEt_3$

The addition of germanium-dichloride, -dibromide or -di-iodide to 1,4-dienes is a versatile method of forming unsaturated 5-membered heterocyclic ring systems (Mironov and Gar, 1965a, 1966; Mazerolles and Manuel, 1966). For example, butadiene and germanium dibromide react exclusively to give the 5-membered ring compound, and 2,3-dimethylbutadiene reacts with germanium di-iodide in a similar way—

$$\text{(butadiene)} + GeBr_2 \longrightarrow \underset{Br_2}{\underset{Ge}{\text{(ring)}}}$$

$$Me\text{-C=C-}Me \text{ (2,3-dimethylbutadiene)} + GeI_2 \longrightarrow Me\underset{I_2}{\underset{Ge}{\diagup}}Me$$

The literature on these reactions, which is somewhat confusing, has been summarized by Nefedov and Manakov (1966). One complicating feature is that trichlorogermane, particularly as its etherate, behaves partly as germanium dichloride in its addition reactions with dienes (Chapter 7).

Six-membered ring systems incorporating two germanium atoms have been obtained from the reaction between germanium di-iodide and acetylene or diphenylacetylene. Again the literature is complicated, since the early papers assumed erroneously that germanium di-iodide added across the triple bond of the acetylene forming a 3-membered ring, and several derivatives were reported based on this structure. It was later shown, mainly by mass-spectroscopic molecular-weight measurements that the 6-membered ring structure is correct (Johnson *et al.*, 1965)—

$$GeI_2 + 2C_2H_2 \xrightarrow{140°C} I_2Ge \underset{}{\overset{}{\bigcirc}} GeI_2$$

$$GeI_2 + PhC \vdots CPh \longrightarrow I_2Ge \underset{Ph\ Ph}{\overset{Ph\ Ph}{\bigcirc}} GeI_2$$

Germanium dichloride reacts in the same way as the di-iodide. This 1,4-digermin ring structure has been further confirmed by several X-ray structural determinations. The carbon—carbon bond length ($1 \cdot 34 \pm 0 \cdot 02$ Å) is very close to that in simple olefins, and the germanium—carbon bond length is $1 \cdot 96$ Å (Bokii and Struchkov, 1967). Much of the confusion in the earlier literature is resolved by Vol'pin *et al.* (1967), who have examined various reactions of this type of compound. The hetero-ring is planar and of quite high stability. The halogen atoms undergo all the usual reactions of organogermanium halides (hydrolysis, halide exchange, alkyl- and aryl-ation) without affecting the ring system. The carbon—carbon double bonds may be reduced or brominated—

These compounds also undergo a Diels–Alder reaction with either 1 or 2 moles of cyclopentadiene—

The germanium di-iodide–acetylene reaction does not give exclusively the 1,4-digermin; benzene-soluble polymeric material is also produced which may be methylated with methylmagnesium iodide to give a polymer having a mean molecular weight of about 33,000 and which is said to be linear.

$$GeI_2 + 2C_2H_2 \longrightarrow (C_2H_2GeI_2)_n \xrightarrow{MeMgI} (C_2H_2GeMe_2)_n$$

2-2'-Dilithiodiphenyl and its derivatives have been used in forming germanium-containing heterocyclics—

Addition of diphenylstannane to diphenyldivinylgermane also produces a cyclic product (Henry and Noltes, 1960)—

Ethylmagnesium bromide, under the right conditions, cleaves a germanium—carbon bond in the germanacyclopentene (Mazerolles and Manuel, 1966)—

A highly stable spiro-germane derived from germanium tetrachloride and the 1,4-dilithio-derivative of tetraphenylbutadiene has been reported (Leavitt et al., 1960)—

In this reaction if the dilithio-compound is added to germanium tetra-chloride or phenyltrichlorogermane, rather than the reverse addition, the chloro-derivatives are formed in 50–70% yield (Curtis, 1967).

The monochloro-compound (3) reacts with butyl-lithium forming a bright-red solution of the lithio-complex (4), which is formally analogous to the cyclopentadienyl ion; but it does not follow that the charge is delocalized. The colour of the solution may not be significant, since the diphenyl derivative (5) also forms an intense purple solution with butyl-lithium which gives a weak e.s.r. signal; hence the main reaction may be addition of butyl-lithium to a carbon—carbon double bond.

The related compounds (6) and (7) have been obtained in which nitrogen is incorporated into the heterocyclic systems (Gilman and Zuech, 1960).

VI. OTHER TYPES

A. Organogermanium carboxylic acids and esters

Compounds of the type R_3GeCO_2H have been isolated in only a few cases, possibly owing to low thermal stability. Carbonation of triphenylgermyl-lithium gives the germyl carboxylic acid as a well crystalline solid (m.p. 189°C) (Brook and Gilman, 1954)—

$$Ph_3GeLi + CO_2 \longrightarrow Ph_3GeCO_2Li \xrightarrow{H_2SO_4} Ph_3GeCO_2H$$

The existence of germyl Grignard reagents was based partly on the isolation of triphenylgermylcarboxylic acid in low yield (Gilman and Zuech, 1961; Glockling and Hooton, 1962b)—

$$Ph_3GeH + CH_2:CHCH_2MgCl \longrightarrow Ph_3GeMgCl \xrightarrow{CO_2} Ph_3GeCO_2H$$

$$GeCl_4 + p\text{-}MeC_6H_4MgCl + Mg \longrightarrow (p\text{-}MeC_6H_4)_3GeMgCl \xrightarrow{CO_2}$$
$$(p\text{-}MeC_6H_4)_3GeCO_2H$$

Thermal decomposition of Ph_3GeCO_2H takes place at its melting point; carbon monoxide is evolved, and the resulting triphenylgermanol is partly esterified by unreacted acid—

$$Ph_3GeCO_2H \xrightarrow{189°} CO + Ph_3GeOH \xrightarrow{Ph_3GeCO_2H} Ph_3GeOCOGePh_3$$

On further heating the ester decomposes, again with loss of carbon monoxide, to give hexaphenyldigermoxane—

$$Ph_3GeOCOGePh_3 \rightarrow (Ph_3Ge)_2O + CO$$

There is evidence that tribenzylgermyl-lithium forms a carboxylic acid, $(PhCH_2)_3GeCO_2H$, that is unstable at room temperature (Cross and Glockling, 1964). An asymmetric lithiogermane has been carbonated with retention of configuration (p. 65). No trialkylgermyl carboxylic acids have been isolated, although permanganate oxidation of $Et_3GeCH:CHBu^n$ is said to yield the crude unstable acid Et_3GeCO_2H (Satge, 1961).

B. Germyl ketones R_3GeCOR'

There are several preparative methods available for α-germyl ketones. Free-radical bromination of a benzylgermane with N-bromosuccinimide followed by hydrolysis of the resulting dibromide has been used in one case (Brook *et al.*, 1960)—

$$Ph_3GeCH_2Ph \xrightarrow{\textit{N}\text{-bromosuccinimide}} Ph_3GeC(Br)_2Ph \xrightarrow{H_2O} Ph_3GeCOPh$$

A more direct method uses triphenylgermyl-lithium and aromatic acid chlorides at low temperature (Allred and Nicholson, 1965)—

$$Ph_3GeLi + ArCOCl \xrightarrow{-50°C} Ph_3GeCOAr + LiCl$$

Some care is needed in carrying out these reactions, since, as with simple organolithium reactions, the carbonyl group will undergo further reaction, so that the carbinol is the final product isolated, and with aliphatic acid chlorides this second reaction is rapid—

$$Ph_3GeCOPh + Ph_3GeLi \rightarrow (Ph_3Ge)_2C(OH)Ph$$

Bistriphenylgermyl ketone is of interest; it forms an orange–pink solid [$(Ph_3Si)_2CO$ is red–violet] that is quite unstable in solution, decomposing to carbon monoxide and hexaphenyldigermane (Brook and Peddle, 1966)—

$$2Ph_3GeLi + HCO_2Me \rightarrow (Ph_3Ge)_2CO$$
$$(Ph_3Ge)_2CO \rightarrow (Ph_3Ge)_2 + CO$$

A further route to α-germyl ketones makes use of the ready hydrolysis of germyl dithianes (Corey *et al.*, 1967; Brook *et al.*, 1967)—

$$R'_3GeCOR$$

The reaction of germyl ketones with alkylidenephosphoranes gives the products of a normal Wittig reaction (Brook and Fieldhouse, 1967)—

In triphenylacetylgermane, $Ph_3GeCOCH_3$, the carbonyl bond length ($1·20 \pm 0·02$ Å) is not significantly different from the value found in ketones. However the germanium—acetyl bond length ($2·011$ Å) is appreciably longer than the germanium—phenyl bond ($1·945$ Å) (Harrison and Trotter, 1968).

The electronic absorption spectra of α-silyl- and α-germyl-ketones are of particular interest in relation to possible d_π–p_π bonding. Replacement of an α-carbon atom by silicon or germanium results in a red shift in the $n \rightarrow \pi^*$

transition of the order of 6000–8000 cm^{-1}, whereas the effect on the position of the $\pi \rightarrow \pi^*$ transition is much smaller (500–1800 cm^{-1}) but in the same direction. These spectroscopic differences are most readily explained in terms of differences in σ-donor bonding between carbon and silicon or germanium. Any d_π–p_π bonding is probably of secondary importance even for silicon, and the similarity in the spectra of silyl- and germyl-ketones is also consistent with the near-absence of π-bonding (Agolini *et al.*, 1968).

C. Fluorocarbon–germanium compounds

The chemistry of fluorocarbon derivatives of metals has been surveyed by Treichel and Stone (1964). Some of the compounds show the most unexpected properties, for example bistrifluoromethylmercury is water soluble.

Trifluoromethyliodogermanes have been prepared by the reaction (Clark and Willis, 1962)—

$$F_3CI + GeI_2 \rightarrow F_3CGeI_3 + (F_3C)_2GeI_2$$

The main product, trifluoromethyltri-iodogermane, is a dense yellow liquid that decomposes above 180°C to the germanium tetrahalides and fluorocarbons—

$$F_3CGeI_3 \xrightarrow{180°C} GeF_4 + GeI_4 + \text{fluorocarbons}$$

It is decomposed by base with the formation of fluoroform.

Perfluoroalkenylgermanes have received some study (Coyle *et al.*, 1961; Stafford and Stone, 1961)—

$$F_2C:CFMgBr \underset{\xrightarrow{Me_2GeCl_2}}{\overset{\xrightarrow{GeCl_4}}{}} \begin{array}{l} (F_2C:CF)_4Ge \\ Me_2Ge(CF:CF_2)_2 \end{array}$$

Addition of triphenylgermyl-lithium to triethylperfluorovinylsilane is followed by spontaneous elimination of lithium fluoride (Seyferth and Wada, 1962)—

$$Et_3SiCF:CF_2 + Ph_3GeLi \longrightarrow \underset{F}{\overset{Et_3Si}{>}}C{=}C\underset{GePh_3}{\overset{F}{<}} + LiF$$

Fully fluorinated arylgermanes are readily prepared either from the lithium aryl and germanium halide or from the metal and iodo-polyfluoroarenes (Tamborski *et al.*, 1965; Chambers and Cunningham, 1965; Fenton *et al.*, 1966; Cohen and Massey, 1968)—

$$4C_6F_5Li + GeCl_4 \xrightarrow{-78°C} (C_6F_5)_4Ge \xleftarrow{325°C} Ge + 4C_6F_5I$$

$$3C_6F_5Li + GeCl_4 \longrightarrow (C_6F_5)_3GeCl$$

(8)

The thermal stability of fluoroarylgermanes is considerable, and bis(octa-fluorobiphenylene)germane (8) undergoes only slight decomposition over several hours at 500°C. Tetrakispentafluorophenylgermane is unaffected by iodine, bromine or iodine chloride under conditions that cleave phenyl—germanium bonds. It is also stable to 5M sodium hydroxide at 20°C. The chloride, $(C_6F_5)_3GeCl$, has been hydrolysed to the germanol, but sodium in hot xylene attacks the carbon—fluorine bonds. The chemical inertness of these compounds is most probably due to the inductive effect of the C_6F_5 groups in inhibiting nucleophilic attack at germanium.

D. Germyl-substituted ferrocene

Metallation of ferrocene by amylsodium followed by reaction with triphenylbromogermane yields a mixture of the mono- and di-triphenylgermyl derivatives (Seyferth *et al.*, 1962a)—

Germyl-substituted ferrocenes, produced in aluminium chloride-catalysed reactions, have been interpreted in terms of the formation of germonium ion species, such as $(Me_2N)_3Ge^+$ and $[(Me_2N)_2Ge-C_5H_4FeC_5H_5]^+$. If bisdimethylaminodichlorogermane and ferrocene are heated with aluminium chloride and the mixture subsequently hydrolysed, two germoxanes are produced—

$$(Me_2N)_2GeCl_2 + \pi\text{-}(C_5H_5)_2Fe \xrightarrow{AlCl_3} \left(\left[\begin{array}{c} C_5H_4 \!-\! GeO \!-\! \\ | \\ Fe \\ | \\ C_5H_5 \end{array} \right]_2 \right)_3 + \left(\left[\begin{array}{c} C_5H_4 \!-\! Ge \!-\! O \\ | \\ Fe \\ | \\ C_5H_5 \end{array} \right]_3 \right)_2$$

Similarly tetrakisdimethylaminogermane yields tris(ferrocenyl)chloroger-mane and its hydrolysis product (Sollott and Peterson, 1967)—

$$(Me_2N)_4Ge + \pi\text{-}(C_5H_5)_2Fe \xrightarrow{AlCl_3} \left[\begin{array}{c} C_5H_4 \!-\! GeCl \\ | \\ Fe \\ | \\ C_5H_5 \end{array} \right]_3 + \left(\left[\begin{array}{c} C_5H_4 \!-\! Ge \!-\! O \\ | \\ Fe \\ | \\ C_5H_5 \end{array} \right]_3 \right)_2$$

VII. REACTIONS

Cleavage reactions, such as halogenation, metallation and redistribution, are discussed in other Chapters. The thermal stability of the tetra-alkyls and tetra-aryls is considerable. For example, tetraethylgermane decomposes over the temperature range 420°–450°C, forming germanium, hydrogen and various hydrocarbons (Geddes and Mack, 1930) and bond energy measurements are consistent with the primary process being cleavage of an ethyl radical (Rabinovich et al., 1963)—

$$Et_4Ge \rightarrow Et_3Ge^{\cdot} + Et^{\cdot}$$

Germanium—carbon bonds are extremely resistant to hydrogenolysis and tetrabenzylgermane is unaffected by hydrogen at 300°C and 300 atm, even in the presence of a palladium catalyst (Glockling and Wilbey, 1968). Only germanacyclanes have been reduced by metal hydrides. Tetraphenyl-germane resists hydrogenation under conditions where tetraphenylsilane is converted into tetracyclohexylsilane (Spialter et al., 1965). This inertness means that any reducible functional groups can be hydrogenated without germanium—carbon bond cleavage, and many reactions of this type have been reported.

Strong oxidizing agents are necessary to convert organogermanes to germanium dioxide (p. 30); by contrast, sulphur cleaves germanium—carbon bonds under quite mild conditions. Arylgermanes are considerably more reactive than the alkyls, especially towards acid and base solvolysis and metallation. Many oxidative reactions have been performed on functional groups, although the low yields suggest that some degradation occurs (Benkeser *et al.*, 1956)—

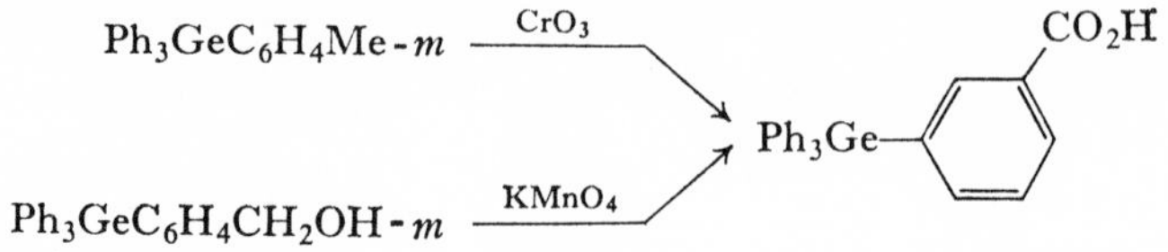

$$Ph_3GeC_6H_4Me\text{-}m \xrightarrow{\quad CrO_3 \quad}$$

$$Ph_3GeC_6H_4CH_2OH\text{-}m \xrightarrow{\quad KMnO_4 \quad} Ph_3Ge\text{—}\underset{}{\bigcirc}\text{—}CO_2H$$

Triethylcyclopentylgermane evidently survives dehydrogenation at 600°C (Nefedov and Manakov, 1963)—

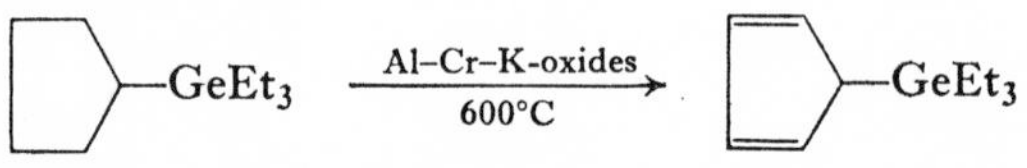

$$\text{(cyclopentyl)}\text{—GeEt}_3 \xrightarrow[600°C]{\text{Al–Cr–K-oxides}} \text{(cyclopentadienyl)}\text{—GeEt}_3$$

Free-radical halogenation of alkylgermanes can be carried out without germanium—carbon bond cleavage—

$$EtGeCl_3 + SO_2Cl_2 \xrightarrow{\;(PhCO)_2O_2\;} \underset{90\%}{CH_2ClCH_2GeCl_3} + \underset{10\%}{CH_3CHClGeCl_3}$$

Photochemical chlorination of ethyltrichlorogermane gives a different ratio of isomers, suggesting that the $GeCl_3$ group has a pronounced orientating effect (Kadina *et al.*, 1966). n-Propyltrichlorogermane is monochlorinated by sulphuryl chloride in the β- and γ-positions, whereas methyltrichlorogermane gives both mono- and di-chloromethyl derivatives (Mironov *et al.*, 1959). Photochlorination of methyltrichlorogermane at 150°C produces a mixture of all three chlorination products (Ponomarenko and Vzenkova, 1957). Decomposition of tetraethylgermane by t-butylbenzoyl peroxide results in abstraction of hydrogen from a β-carbon atom and formation of the radical dimer in low yield (Vyazankin *et al.*, 1965)—

$$Et_4Ge + Me_3COOCOPh \rightarrow Et_3Ge(CH_2CH_2)_2GeEt_3$$

Metathetical reactions and even Wurtz condensation have been carried out without affecting germanium—carbon bonds, as in the following illustrations (Bott *et al.*, 1962; Mironov and Kravchenko, 1963)—

$$Et_3Ge(CH_2)_4Br \xrightarrow{\;KCN\;} Et_3Ge(CH_2)_4CN$$

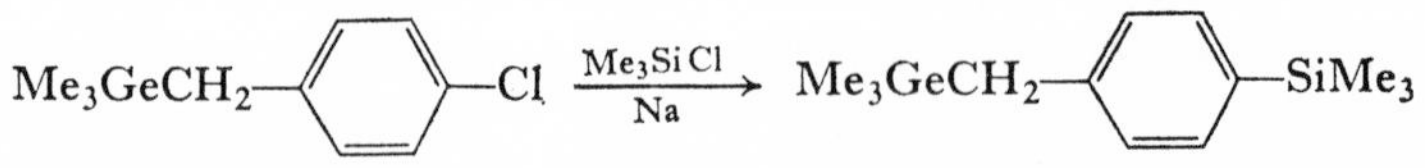

$$Me_3GeCH_2\text{—}\underset{}{\bigcirc}\text{—Cl} \xrightarrow[Na]{Me_3SiCl} Me_3GeCH_2\text{—}\underset{}{\bigcirc}\text{—SiMe}_3$$

Some simple reactions, such as the hydrolysis of triethylgermyl acetic ester, lead to germanium—carbon bond cleavage (Rijkens *et al.*, 1964)—

$$Et_3GeCH_2CO_2Et \xrightarrow{\;H_2O\;} (Et_3Ge)_2O + CH_3CO_2Et$$

Grignard derivatives of haloalkylgermanes are readily prepared and can be used in a wide variety of reactions—

$$Me_3GeCH_2Cl \rightarrow Me_3GeCH_2MgCl \xrightarrow{BrCH_2CH:CH_2} Me_3Ge(CH_2)_2CH:CH_2$$

Tetrabenzylgermane has been sulphonated in the aromatic rings using fuming sulphuric acid, the composition of the product being $(HOSO_2-C_6H_4CH_2)_4Ge$ (Orndorff et al., 1927).

Organogermanes of the type $R_3GeCH_2CH_2X$ commonly eliminate alkene under mild conditions (Mironov and Dzhurinskaya, 1963; Vol'pin et al., 1967)—

$$ClCH_2CH_2GeCl_3 \xrightarrow{MeMgBr} ClCH_2CH_2GeMe_3 \longrightarrow Me_3GeCl + C_2H_4$$

$$\downarrow MeMgBr$$

$$Me_4Ge$$

$$MeCH(Cl)CH_2GeCl_3 \xrightarrow{MeMgBr} Me_4Ge + C_3H_6$$

The tendency towards β-elimination reactions of this type increases in the order Si < Ge < Sn < Pb, which suggests that the reactions involve penta-co-ordinate intermediates—

$$Me_3Ge \overset{Cl}{\underset{CH_2}{\diagup\diagdown}} CH_2 \longrightarrow Me_3GeCl + C_2H_4$$

Acid and base solvolysis

The acid solvolysis of aryl-germanium bonds by perchloric acid in aqueous ethanol has been the subject of several investigations (Eaborn and Pande, 1960a, 1961; Bott et al., 1962, 1964a, b). Comparative measurements on the solvolysis of such compounds as $4\text{-}MeOC_6H_4MR_3$, where M = silicon, germanium, tin and lead, reveal a large increase in rate between germanium and tin (relative reactivities: Si, 1; Ge, 36; Sn, $3 \cdot 5 \times 10^5$; Pb, 2×10^8). For silicon and germanium the mechanism is considered to involve primary proton attack at carbon, whereas with tin and lead rapid co-ordination of water giving a pentaco-ordinate intermediate may precede proton attack at carbon. There is of course a progressive weakening of the metal—carbon bond in this series.

The cleavage of aryl—germanium bonds by perchloric acid is enhanced by para-alkyl substituents according to the Baker–Nathan order Me > Et > Pr^i > Bu^t. Activation by meta-alkyl groups follows the order of inductive effects (Bu^t > Me). The relative effects of various other substituents have been noted; *para*-halogen atoms deactivate in the order $I \sim Br > Cl > F$, and the methoxy group activates more from the *para* than from the *ortho* position. Trimethylgermyl groups appear to be more strongly electron donating than trimethylsilyl. In substituted pyridines of the type $2\text{-Me}_3MC_5H_4N$, the order of electron release is similar, $Me_3Si \sim Me_3Ge < Me_3Sn$ (Anderson *et al.*, 1968).

Most organogermanes are unaffected by aqueous caustic alkalis, and there is no report on their fusion with sodium hydroxide. However, benzyl-, diphenylmethyl- and fluorenyl-germanes are susceptible to base solvolysis; these reactions are best regarded as involving nucleophilic attack at germanium. Comparative rate studies on silanes, germanes and stannanes reveal that the salient difference is a large decrease in rate between benzyl-silanes and -germanes and a slight increase with -stannanes (Bott *et al.*, 1963a). Relative reactivities for $R_3MCH_2C_6H_4Cl\text{-}m$ are: Ge, 1×10^{-3}; Si, $1\cdot0$; Sn, 17. Reaction rates also vary with the structure of the organic group cleaved in the order: fluorenyl $> Ph_3C > Ph_2CH > PhCH_2$.

Nucleophilic attack is thus easier on silicon than germanium, and this has been ascribed to increased nuclear shielding with germanium without appreciable increase in atomic size. 9-Fluorenyl groups are quantitatively cleaved from germanium by refluxing for 30 min with weakly alkaline aqueous ethanol.

Kinetic studies on the cleavage of allyl—germanium bonds by mercuric chloride in acetonitrile solution are consistent with a bimolecular nucleophilic mechanism—

$$Et_3GeCH_2CH:CH_2 + HgCl_2 \rightarrow Et_3GeCl + CH_2:CHCH_2HgCl$$

Relative rates for similar silanes and stannanes ($Et_3Sn \gg Et_3Ge > Et_3Si$) reflect the electron-release properties of the Et_3M groups. The base-catalysed hydrolysis of allylic germanium compounds is also consistent with an S_N2 mechanism, in which the rate-determining step is attack on the metal with expulsion of a carbanion (Roberts, 1968; Roberts and Kaissi, 1968). Trifluoroacetic acid will cleave one phenyl group from tetraphenyl-germane (Sartori and Weidenbruch, 1967).

VIII. POLYMERIC ORGANOGERMANES

Several organogermanes containing olefinic groups have been polymerized by using free radical initiators. Vinyl-, allyl- and styryl-germanes form oligomers with 3–5 monomer units in the chain, whereas copolymers with styrene or methylmethacrylate yield solid products in which hydrocarbon units predominate. The copolymers generally soften at higher temperatures than polystyrene or polymethylmethacrylate (Korshak *et al.*,

1959; Brinckman and Stone, 1959; Noltes *et al.*, 1960; Kolesnikov *et al.*, 1961). Triethylgermyl methacrylate produces a solid polymer having a molecular weight of about 200,000, and triphenylgermyl methacrylate has been block and solution polymerized to yield a hard clear thermoplastic and a white powder, respectively (Florinskii, 1962).

Diphenylgermane may be polymerized by di-t-butylperoxide, producing a dark brown amorphous solid that has been formulated as $[Ph(MeC_6H_4)\text{-}GeCH_2]_n$, with a molecular weight of about 1840 (Sosin *et al.*, 1964). Diphenylstannane and diphenyldistyrylgermane produce a polymer of fairly high molecular weight, formed by addition of SnH to the olefinic bond (Noltes and Van der Kerk, 1961)—

$$Ph_2SnH_2 + Ph_2Ge(C_6H_4CH:CH_2\text{-}p)_2$$

4 | ORGANOGERMANIUM HYDRIDES

Alkyl- and aryl-germanium hydrides of all types show an enhanced thermal stability and decreased sensitivity to oxidation over the parent hydrides. The general order of stability appears to be $R_3GeH > R_2GeH_2 > RGeH_3$. Few bond-energy measurements have been reported, but it seems clear that the germanium—hydrogen bond strength is intermediate between silicon—hydrogen and tin—hydrogen in similar organo-derivatives. The reactivity of organogermanium hydrides shows some striking differences from silicon and tin analogues, and these probably relate to differences in bond polarity and availability of d orbitals. For example, both trialkylsilanes and trialkylstannanes are considerably more sensitive to acid or base solvolysis than trialkylgermanes. Several trialkylgermanes have even been prepared in strongly acid solution. The halogenation of organogermanium hydrides and their reaction with pseudohalides is discussed in Chapter 7, and reactions with transition-metal complexes and other organometallic compounds such as mercury dialkyls in Chapter 8.

I. PREPARATIVE METHODS

Organogermanium halides are readily reduced by lithium aluminium hydride. This is a general reaction with few complications, and has been extensively employed in the preparation of mono-, di- and tri-hydrides. The most common solvent is ether, but tetrahydrofuran, monoglyme, diglyme and di-n-butyl ether have also been used, depending on the volatility of the product. Organogermanium hydrides may be distilled directly from the reaction medium or isolated after a hydrolysis stage. Lithium hydride in ethereal solvents has been used in the same way (Ponomarenko *et al.*, 1958). Sodium borohydride is also a satisfactory reducing agent and may be used in ethereal or dilute aqueous solution (Satge, 1961; Griffiths, 1963b). Reductive cleavage of germanium—oxygen bonds has not been extensively studied, but an asymmetric menthyl compound has been reduced in high yield by lithium aluminium hydride, with retention of configuration (p. 64).

The use of zinc amalgam in aqueous acid for preparing organogermanium hydrides is of greatest interest in demonstrating their stability under strongly acid conditions (12M hydrochloric acid), and attempts to prepare silicon and tin hydrides by similar procedures have been unsuccessful (West,

1953). The yields are poor and metal hydride reduction of halides is far superior from a preparative point of view—

$$Ph_3GeBr \xrightarrow[\text{HCl}]{\text{Zn/Hg}} Ph_3GeH \quad (15\%)$$

$$(Me_2GeS)_3 \xrightarrow[\text{HCl}]{\text{Zn/Hg}} Me_2GeH_2 \ (5\%)$$

Methyltrichlorogermane failed to produce any methylgermane by this method. The hydrolysis of metal complexes, which may be illustrated by—

$$Ph_3GeNa + H_2O \rightarrow Ph_3GeH + NaOH$$

is not of general preparative value, but has been used to provide evidence that di-alkali metal derivatives, such as Ph_2GeNa_2, are formed in cleavage reactions (Neumann and Kuehlein, 1967a)—

$$(Ph_2Ge)_4 \xrightarrow{\text{Na, } C_{10}H_8} Ph_2GeNa_2 \xrightarrow{\text{H}_2\text{O}} Ph_2GeH_2$$

It also provides evidence for the formation of germyl Grignard reagents (Chapter 8)—

$$R_3GeX \xrightarrow{\text{Mg}} R_3GeMgX \xrightarrow{\text{H}_2\text{O}} R_3GeH$$

Organogermanium hydrides have been obtained by the partial alkylation of di- or tri-hydrides by Grignard reagents (Satge, 1961)—

$$R_2GeH_2 \xrightarrow[\text{THF}]{\text{R'MgX}} R_2R'GeH + (MgH_2)_n + MgX_2$$

$$RGeH_3 \xrightarrow[\text{THF}]{\text{R'MgX}} RR'GeH_2 + RR'_2GeH + (MgH_2)_n + MgX_2$$

This provides a route to germanium hydrides having different organic groups attached to germanium. Di-n-butylallylgermane has been obtained in 35% yield and di-n-butylbenzylgermane in 28%. Diazo-compounds have been used in the partial alkylation of organogermanium di- and tri-hydrides (p. 91).

Functionally substituted mercury alkyls are sufficiently reactive to alkylate germanium hydrides, and with dihydrides monoalkylation may be achieved (Baukov and Lutsenko, 1964)—

$$Bu^n_2GeH_2 + Hg(CH_2CO_2Me)_2 \rightarrow Bu^n_2Ge(H)CH_2CO_2Me + CH_3CO_2Me + Hg$$

A. Hydrogermylation of alkenes and alkynes

The addition of germane to ethylene is a complex reaction that produces only low yields of ethylgermane (Mackay and Watt, 1968)—

$$C_2H_4 + GeH_4 \xrightarrow{\text{AlCl}_3} EtGeH_3$$

Organogermanium di- and tri-hydrides add to carbon—carbon double and triple bonds, but rather less readily than the halohydrides discussed in

Chapter 7. Some of these reactions may be controlled so as to yield a germanium hydride, whereas others proceed to completion, giving a tetraorganogermane—

$$R_2GeH_2 \xrightarrow{R'CH:CH_2} R_2Ge(H)CH_2CH_2R'$$

$$\downarrow R'CH:CH_2$$

$$R_2Ge(CH_2CH_2R')_2$$

$$\text{n-}C_7H_{15}GeH_3 + 3CH_2:CHO(CH_2)_3Me \longrightarrow \text{n-}C_7H_{15}Ge[CH_2CH_2O(CH_2)_3Me]_3$$

B. Organodigermanium hydrides

Partially substituted organodigermanes of the type $(R_2GeH)_2$ have been obtained as low yield by products from various reactions (Cross and Glockling, 1964; Carrick and Glockling, 1966). The only systematic study of these compounds is by Mackay et al. (1968b) who effected low-temperature alkylation of the unstable digermyl iodide—

$$Ge_2H_5I \xrightarrow{RMgX} H_3GeGeH_2R \quad (R = Me, Et)$$

These mono-alkylated digermanes are much more stable than the digermyl halides and their spectroscopic properties have been examined in some detail.

II. ORGANOGERMANIUM DEUTERIDES

These are readily obtained by metal deuteride reduction of germanium halides (Ponomarenko et al., 1958, 1961; Griffiths, 1963b; Mackay and Watt, 1967). They are also formed by the deuterolysis of germyl-alkali metal or -Grignard derivatives (Cross and Glockling, 1964, 1965a; Carrick and Glockling, 1966)—

$$GeCl_4 + Pr^iMgCl \longrightarrow Pr^i_3GeMgCl \xrightarrow{D_2O} Pr^i_3GeD$$

$$(PhCH_2)_4Ge \longrightarrow (PhCH_2)_3GeLi + (PhCH_2)_2GeLi_2$$

$$\downarrow D_2O$$

$$(PhCH_2)_3GeD + (PhCH_2)_2GeD_2$$

$$Ph_2GeH_2 \xrightarrow{Bu^nLi} Ph_2(Bu^n)GeLi \xrightarrow{D_2O} Ph_2(Bu^n)GeD$$

There seems to be no example of direct hydrogen–deuterium exchange being applied to the formation of deuterides.

III. REACTIONS

The thermal stability of organogermanium hydrides is considerable, but some literature reports confuse aerial oxidation with thermal stability. Thus diphenylgermane is said to decompose slowly at room temperature, but this is almost certainly an oxidative decomposition, since in an evacuated sealed tube it is unchanged over 8 hr at 250°C, even when irradiated with u.v. light. Between 280° and 340°C, decomposition occurs with the formation of a germanium mirror, hydrogen, tetraphenylgermane and benzene. The formation of these products suggests a series of radical reactions, with the possible intermediate formation of diphenylgermanium (Glockling and Hooton, 1963)—

$$Ph_2GeH_2 \rightarrow H_2 + C_6H_6 + Ge + Ph_2Ge$$
$$2Ph_2Ge \rightarrow Ph_4Ge + Ge$$

Triphenylgermane disproportionates on heating (Johnson and Harris, 1950)—

$$2Ph_3GeH \xrightarrow{300°C} Ph_4Ge + Ph_2GeH_2$$

However tribenzylgermane shows only superficial decomposition at 310°C in the total absence of air. At 400°C, extensive, probably radical, decomposition occurs giving toluene, bibenzyl, *trans*-stilbene, hydrogen and a brown polymer having the approximate composition $Ge_3(CH_2Ph)_2$ (Cross and Glockling, 1964). A similar stability is reported for tricyclohexylgermane which, between 360°–400°C gives hydrogen, benzene, cyclohexene, cyclohexane, condensed hydrocarbons and germanium (Petukhov et al., 1966).

Organogermanium hydrides are moderately strong reducing agents. They are much less sensitive to aerial oxidation than organotin hydrides, but many undergo slow oxidation on exposure to air, the products being organogermanium oxides or hydroxides. Sensitivity to oxidation decreases in the order $RGeH_3 > R_2GeH_2 > R_3GeH$. If a current of air is passed through n-octylgermane, the compound rapidly becomes viscous and the germanoic anhydride may finally be isolated as a white powder (Satge, 1961)—

$$C_8H_{17}GeH_3 \xrightarrow{O_2} [(C_8H_{17}GeO)_2O]_n$$

Di-isopropylgermane is slowly oxidized in air (Anderson, 1956a) whilst tricyclohexylgermane, if refluxed in carbon tetrachloride with continuous passage of air through the solution, is converted into tricyclohexylgermanol (Johnson and Nebergall, 1949).

Strong chemical oxidants, such as potassium permanganate in acetone, oxidize all types of germanium–hydrogen compounds to the appropriate

oxides and many other transition-metal salts are reduced by organogermanes (Anderson, 1957). Reduction occurs if the change in oxidation state
of the metal is between about -0.06 and -2.0 V. Thus platinum(IV),
gold(III) and palladium(II) are reduced to the metal by triethylgermane,
whereas mercury(II) is reduced to mercury(I) if it is present in excess,
otherwise mercury is formed. The reduction of other metal salts depends on
the experimental conditions $[Cu^{II} \rightarrow Cu^{I}; Ti^{IV} \rightarrow Ti^{III} + Ti^{II}; V^{V} \rightarrow V^{IV}$
$+ V^{III}; Mn^{VII} \rightarrow Mn^{IV}; Cr^{VI} \rightarrow Cr^{III}]$. Various halogen-containing organic compounds have been reduced by trialkylgermanes. For example,
carbon tetrachloride gives chloroform, and benzoyl chloride is converted
into benzaldehyde. With alkylgermanium trihydrides reactions of this type
are rapid and exothermic.

Organogermanium hydrides are quite remarkably stable to mineral acids
and caustic alkalis, in contrast to both organo-silanes and -stannanes. For
example, dialkylgermanes have been recovered unchanged after 4 hr refluxing with 60% hydrobromic acid. The following reactions also illustrate this
high stability (Satge, 1961; Massol and Satge, 1966)—

$$2PhGe(H)_2Cl \xrightarrow{10\% \ NH_4OH} (PhGeH_2)_2O$$

$$(Ph_2GeH)_2O \xrightarrow{aq. \ HBr} 2Ph_2Ge(H)Br$$

$$2Me_2Ge(H)X \xrightarrow{NH_3, \ Et_2O} (Me_2GeH)_2NH$$

Organogermanium hydrides, unlike tin analogues, are not catalytically
decomposed by amines, but concentrated sulphuric acid and benzene-
sulphonic acid decompose trialkylgermanes with liberation of hydrogen—

$$2R_3GeH \xrightarrow{conc. \ H_2SO_4} (R_3Ge)_2SO_4 + 2H_2$$

Trifluoroacetic acid converts triethylgermane into the ester, but acetic acid
does not react. Silver salts of carboxylic acids have been used in the formation of germyl esters, and the reaction between organogermanium hydrides
and silver pseudohalides is discussed in Chapter 7 (Anderson, 1957)—

$$Et_3GeH + F_3CCO_2H \rightarrow Et_3GeOCOCF_3 + H_2$$
$$Pr^i_2GeH_2 + 2AgOAc \rightarrow Pr^i_2Ge(OAc)_2 + H_2 + 2Ag$$

The rate of alkaline solvolysis of triorganogermanes is too slow to measure
in hot 0.5% alcoholic potassium hydroxide (Schott and Harzdorf, 1961),
but by contrast hexylgermane liberates hydrogen with alcoholic potassium
hydroxide (Satge, 1961)—

$$n-C_6H_{13}GeH_3 \xrightarrow{KOH, \ EtOH} [(n-C_6H_{13}GeO)_2O]_n + H_2$$

Organogermanium hydrides are quite stable in alcohol solution, but if
heated in the presence of copper powder they decompose with evolution of

hydrogen and formation of an alkoxide. Glycols and phenols react even more readily, and aldehydes, in the presence of platinum or copper powder, form alkoxides by addition of germanium–hydrogen to the carbonyl group (Lesbre and Satge, 1962).

The reaction of triphenylgermane with ethereal n-butyl-lithium has been briefly referred to in Chapter 3 as providing the most satisfactory preparative method for triphenylgermyl-lithium; it does in addition produce triphenylbutylgermane in low yield (Cross and Glockling, 1965a). Metallation reactions are also of interest in relation to the behaviour of triphenylsilane, which is alkylated in high yield rather than metallated (Gilman and Melvin, 1949)—

$$Ph_3SiH + RLi \rightarrow Ph_3SiR + LiH$$

This difference has been ascribed to germanium—hydrogen being less polar in the sense $\overset{\delta+}{Ge}$—$\overset{\delta-}{H}$ than the silicon—hydrogen bond.

Closer examination of these reactions shows that they do not proceed exclusively in one direction; both metallation and alkyl- or aryl-ation occur, and the relative extent of each is dependent on the experimental conditions. If triphenylgermane is added to an excess of phenyl-lithium in refluxing ether, tetraphenylgermane may be isolated in 70% yield (Johnson and Harris, 1950). However reversing the order of addition results in the formation of hexaphenyldigermane, and if only one equivalent of phenyl-lithium is used triphenylgermyl-lithium is formed in about 90% yield (Gilman and Gerow, 1956a)—

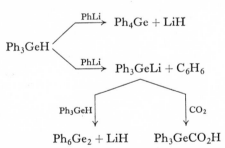

Methyl-lithium is much less reactive towards triphenylgermane than either phenyl- or butyl-lithium, and in refluxing ether some methylation product is formed—

$$Ph_3GeH + MeLi \xrightarrow[\text{2. HCl}]{\text{1. } CO_2} Ph_3GeCO_2H + Ph_3GeMe$$
$$\phantom{Ph_3GeH + MeLi \xrightarrow[\text{2. HCl}]{\text{1. } CO_2}} 70\% \qquad\quad 9\%$$

Lithiation of asymmetric germanes, $R_1R_2R_3Ge^*H$, proceeds with retention of configuration (Chapter 3). The reaction between tribenzylgermane and

4

benzyl-lithium reveals an even more complex situation in which metallation, benzylation and cleavage of a benzyl group are all observed—

$$(PhCH_2)_3GeH + PhCH_2Li \rightarrow$$
$$(PhCH_2)_4Ge + LiH + (PhCH_2)_3GeLi + (PhCH_2)_2GeLi_2$$

$$(PhCH_2)_3GeLi + (PhCH_2)_2GeLi_2 \xrightarrow{\text{MeI}} (PhCH_2)_3GeMe + (PhCH_2)_2GeMe_2$$

In the same way diphenylgermane and n-butyl-lithium react by alkylation, metallation and elimination of lithium hydride.

$$Ph_2GeH_2 + 2BuLi$$

$$\downarrow$$

$$Ph_2GeLi_2 + Ph_2Ge(Bu)Li + Ph_2GeBu_2 + Ph_2(Li)GeGe(Li)Ph_2$$

$$\downarrow \text{EtBr}$$

$$Ph_2GeEt_2 + Ph_2Ge(Bu)Et + Ph_2GeBu_2 + Ph_2(Et)GeGe(Et)Ph_2$$
$$\quad 2\% \qquad\quad 20\% \qquad\quad 12\% \qquad\qquad 28\%$$

Alkylgermanes combine with sodium in liquid ammonia or ethylamine solution with discharge of the blue colour and evolution of hydrogen (Glarum and Kraus, 1950)—

$$RGeH_3 \xrightarrow{\text{Na, NH}_3} RGe(H)_2Na + \tfrac{1}{2}H_2$$

A slow secondary ammonolysis reaction has been postulated—

$$RGe(H)_2Na \xrightarrow{\text{NH}_3} RGe(H)_2NH_2 + H_2 + NaNH_2$$

Ethylgermane, with lithium in ethylamine, evolves hydrogen and ethane, the total hydrogen corresponding to the formation of EtGeLi—

$$EtGeH_3 \xrightarrow[\text{EtNH}_2]{\text{Li}} EtGe(H)_2Li \longrightarrow EtGeLi + H_2$$

The presence of metallated germanium hydrides in these solutions has been firmly established by alkylation reactions—

$$EtGeH_3 \xrightarrow[\text{EtNH}_2]{\text{Li}} EtGe(H)_2Li \xrightarrow{\text{i-C}_5\text{H}_{11}\text{Br}} Et(i-C_5H_{11})GeH_2 \xrightarrow[\text{EtNH}_2]{\text{Li}}$$

$$Et(i-C_5H_{11})Ge(H)Li \xrightarrow{\text{EtI}} Et_2(i-C_5H_{11})GeH$$

The relative acidities of various germanium hydrides have been investigated by their p.m.r. spectra in liquid ammonia (Birchall and Jolly, 1966): $GeH_4 > Ph_3GeH > MeGeH_3 > Me_2GeH_2$. Substitution of hydrogen by methyl or ethyl groups has the expected effect of reducing acidity. Triphenylgermane seems rather anomalous, unless some degree of π-interaction between occupied ring orbitals and the empty $4d$ orbitals of germanium is invoked to account for its weakness relative to germane.

Diazomethane is effective in the partial methylation of organogermanium hydrides (Satge and Riviere, 1966), but the reactions are slow and incomplete, even with irradiation or the addition of copper powder. For example, diphenylgermane with excess of ethereal diazomethane reacts only partially after 6 hr irradiation—

$$Ph_2GeH_2 \xrightarrow[h\nu]{CH_2N_2} \underset{43\%}{Ph_2Ge(H)Me} + \underset{5\%}{Ph_2GeMe_2}$$

The greater ease of monomethylation is brought out by the reaction of phenylgermane with excess of diazomethane—

$$PhGeH_3 \xrightarrow[h\nu]{CH_2N_2} \underset{59\%}{Ph(Me)GeH_2} + \underset{14\%}{Ph(Me)_2GeH}$$

Triphenylgermane does not react with diazomethane, but tributylgermane will combine with several substituted diazomethanes, such as diazoacetone, diazoacetic ester and diazoacetophenone, perhaps because a higher reaction temperature can be obtained (Satge, 1961)—

$$Bu^n_3GeH + N_2CHCO_2Et \xrightarrow{Cu,\ 50°C} Bu^n_3GeCH_2CO_2Et + N_2$$

Dibutylgermane also combines with these diazo-compounds, but a mixture of products results—

$$Bu^n_2GeH_2 + N_2CHCO_2Et \rightarrow Bu^n_2Ge(H)CH_2CO_2Et + Bu^n_2Ge(CH_2CO_2Et)_2$$

Triphenyl- and triethyl-germanes both undergo a carbene insertion reaction (Seyferth et al., 1967a)—

$$Ph_3GeH + PhHgCCl_2Br \rightarrow Ph_3GeCHCl_2 + PhHgBr$$

Silanes behave in the same way, but combine more slowly, whereas tributylstannane reacts mainly by reduction—

$$Bu^n_3SnH + PhHgCBr_3 \rightarrow Bu^n_3SnBr + PhHgCHBr_2$$

The mechanism may involve a pentaco-ordinate transition intermediate formed by attack of the free carbene on the metal, and this is compatible with the observation that asymmetric silanes, $R_1R_2R_3Si{*}H$ insert dichlorocarbene with retention of configuration.

Hydrogermylation of alkenes and alkynes

All types of organogermanium hydrides add to carbon—carbon double and triple bonds, and they appear rather more reactive than similar silanes and stannanes. The reaction conditions vary considerably with the structure of the alkene or alkyne, but most combine when the reactants are heated to about 100°C, often without the presence of a catalyst. Terminal olefins of the type $RCH:CH_2$ react primarily by an anti-Markownikoff addition, although Satge (1961) states that in the addition of trialkylgermanes to acrylonitrile a small proportion of branched chain product is formed—

$$R_3GeH + CH_2:CHCN \xrightarrow{\text{reflux}} R_3GeCH_2CH_2CN + \text{trace } R_3GeCH(CN)CH_3$$

The yields reported in reactions of this type are commonly about 50% (Gilman and Gerow, 1957b; Fuchs and Gilman, 1957; Henry and Downey, 1961; Mikhailov et al., 1965; Mironov and Gar, 1966)—

$$Bu^n_3GeH + CH_2:CHCO_2Me \rightarrow Bu^n_3GeCH_2CH_2CO_2Me \ (40\%)$$
$$Ph_3GeH + PhCH:CH_2 \rightarrow Ph_3Ge(CH_2)_2Ph \qquad (40\%)$$

In a few cases, the anti-Markownikoff addition has been convincingly demonstrated by an alternative synthesis—

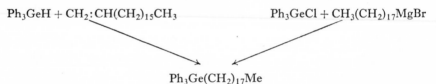

$$Ph_3GeH + CH_2:CH(CH_2)_{15}CH_3 \qquad\qquad Ph_3GeCl + CH_3(CH_2)_{17}MgBr$$

$$Ph_3Ge(CH_2)_{17}Me$$

Chloroplatinic acid has been used as a catalyst for the hydrogermylation of alkenes and alkynes, although its use may not be essential (Dzhurinskaya et al., 1961; Satge, 1961). The detailed mechanism of the catalysed reaction is obscure, but it is commonly assumed that platinum(IV) is reduced and forms a reactive π-complex with the alkene or alkyne—

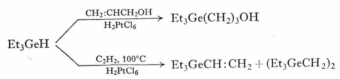

$$Et_3GeH \begin{cases} \xrightarrow[H_2PtCl_6]{CH_2:CHCH_2OH} Et_3Ge(CH_2)_3OH \\ \\ \xrightarrow[H_2PtCl_6]{C_2H_2,\ 100°C} Et_3GeCH:CH_2 + (Et_3GeCH_2)_2 \end{cases}$$

Where a reactive functional group is present, competing reactions may occur as, for example, with allyl bromide in the presence of chloroplatinic acid (Mironov et al., 1962a)

$$Et_3GeH + CH_2:CHCH_2Br \begin{cases} \longrightarrow Et_3GeBr + C_3H_6 \\ \longrightarrow Et_3Ge(CH_2)_3Br \\ \longrightarrow Et_3GeCH_2CH:CH_2 + HBr \end{cases}$$

Cyclopentene and cyclohexene combine as readily with germanium hydrides as do terminal olefins (Mironov et al., 1962b).

Trialkyl- and triaryl-germanes react with acetylene and even more readily with monosubstituted acetylenes in two stages. Since the rate of addition of the germane to a triple bond is considerably greater than the rate of the second stage, these reactions can frequently be used selectively—

$$\text{Bu}^n{}_3\text{GeH} + \text{PhC} \vdots \text{CH} \xrightarrow[20°]{\text{H}_2\text{PtCl}_6} \text{Bu}^n{}_3\text{GeCH} \vdots \text{CHPh} \quad (100\%)$$

Further addition of trialkylgermane produces a 1,2-digermylethane rather than the 1,1-derivative. Many more complicated unsaturated compounds have been examined in connection with hydrogermylation. In compounds containing both alkene and alkyne groups, it is always the alkyne that is the more reactive, even when the alkene is terminal (Gverdtsiteli *et al.*, 1964). Non terminal alkynes will hydrogermylate with a catalyst present. Trimethylgermane adds to cyclopentadiene in the presence of chloroplatinic acid to produce a complexity of products of which two have been characterized (Fish and Kuivila, 1966)—

$$\text{Me}_3\text{GeH} + \begin{array}{c}\bigcirc\end{array} \xrightarrow{\text{H}_2\text{PtCl}_6} \begin{array}{c}\bigcirc\end{array}\!\!-\text{GeMe}_3 + \begin{array}{c}\bigcirc\end{array}\!\!-\text{GeMe}_3$$

Diorganogermanes have been added to conjugated and non-conjugated dienes and to diynes. Both types of reaction lead to polymeric products, although the addition of dibutylgermane to diethyldivinylgermane also produces a 6-membered germanium heterocyclic compound (Mazerolles, 1962)—

$$\text{Bu}^n{}_2\text{GeH}_2 + \text{Bu}^n{}_2\text{Ge(CH} \vdots \text{CH}_2)_2 \longrightarrow \text{polymer} + \text{Bu}^n{}_2\text{Ge} \underset{\diagdown\diagup}{\overset{\diagup\diagdown}{}} \text{GeBu}^n{}_2$$

The reaction between n-heptylgermane and hept-1-yne illustrates the greater reactivity of the alkyne group—

$$\text{n-C}_7\text{H}_{15}\text{GeH}_3 + 3\text{Me(CH}_2)_4\text{C} \vdots \text{CH} \xrightarrow{200°\text{C}} \text{n-C}_7\text{H}_{15}\text{Ge[CH} \vdots \text{CH(CH}_2)_4\text{Me]}_3$$

5 | ORGANOGERMANIUM–NITROGEN AND –PHOSPHORUS COMPOUNDS

I. GERMANIUM–NITROGEN COMPOUNDS

The earliest systematic studies on germanium–nitrogen systems were reported by Thomas and Southwood (1931), and the subject was further developed by Kraus and Flood (1932) and Schwartz and Reinhardt (1932). These investigations were concerned with the reactions between germanium halides and ammonia or amines under anhydrous conditions, sometimes in the presence of alkali metals. The relatively high reactivity of germanium–nitrogen systems towards protic reagents has revived interest in the chemistry of this class of compound; four main types have been studied:

(i) Co-ordination complexes, e.g., $R_3Ge(X)NH_3$, $R_3Ge(X)NR_3'$

(ii) Primary-, secondary- and tertiary-germylamines, e.g., R_3GeNH_2, $(R_3Ge)_2NH$, $(R_3Ge)_3N$, $(R_3Ge)_2NR'$, $R_3GeN(Ph)NHPh$, $R_3GeN(Ph)CO_2R'$

(iii) Cyclic or polymeric compounds, e.g., $(R_2GeNH)_n$, $(R_2GeNR')_3$, $(RGeN)_n$

Organogermanium halides react with anhydrous ammonia and amines in essentially the same way as germanium tetrahalides (Chapter 2). The products isolated depend on the stoicheiometry and on the reaction conditions. Few co-ordination complexes have been isolated; their composition corresponds to pentaco-ordinate species, but no structural evidence has yet been reported. Trimethylbromogermane, for example, forms 1:1 adducts with ammonia, methylamine, dimethylamine and trimethylamine, but those with a reactive hydrogen atom can be induced to eliminate the quaternary ammonium salt, forming a secondary or tertiary germylamine (Kraus and Flood, 1932)—

$$Me_3GeX \xrightarrow{NH_3} Me_3Ge(X)NH_3 \xrightarrow{NH_3} (Me_3Ge)_2NH + NH_4X$$

In ethereal solution, at low temperatures, this reaction has been controlled, giving the secondary and tertiary germylamines (Satge and Baudet, 1966)—

$$Me_3GeCl \xrightarrow[-60°C]{NH_3,\ Et_2O} (Me_3Ge)_2NH + (Me_3Ge)_3N$$
$$50\% \qquad\quad 32\%$$

94

Alkylchlorogermanium hydrides behave similarly (Massol and Satge, 1966)—

$$R_2Ge(H)X \xrightarrow{NH_3} [R_2GeH]_2NH + (R_2GeH)_3N + NH_4Cl$$

$$RGe(H)_2X \xrightarrow{NH_3} (RGeH_2)_3N + NH_4Cl$$

Dimethyldichlorogermane and ammonia have yielded a partial substitution product that may be converted into tris(trimethylgermyl)amine by reaction with methyl-lithium. Trisgermylamines have also been obtained via lithiogermylamides (Ruidisch and Schmidt, 1964; Pflugmacher and Hirsch, 1968)—

$$Me_2GeCl_2 \xrightarrow{NH_3} (Me_2GeCl)_3N \xrightarrow{MeLi} (Me_3Ge)_3N$$

$$(R_3Ge)_2NLi \xrightarrow{R_3GeBr} (R_3Ge)_3N$$

By contrast, dibutyldichlorogermane with excess of sodium in liquid ammonia gives the secondary germylamine as a cyclic trimer (Rijkens et al., 1965; Luijten et al., 1965) together with polymeric material—

$$Bu^n{}_2GeCl_2 \xrightarrow{Na/NH_3} Bu^n{}_2Ge \begin{array}{c} Bu^n{}_2 \\ NH-Ge \\ \diagup \qquad \diagdown \\ \qquad\qquad NH \\ \diagdown \qquad \diagup \\ HN-Ge \\ Bu^n{}_2 \end{array}$$

Similarly, dimethyldichlorogermane and methylamine form a cyclic trimer, $(Me_2GeNMe)_3$, in ethereal solution. Organogermanium trihalides and ammonia yield polymeric compounds that have been scarcely studied (Flood, 1933)—

$$EtGeI_3 \xrightarrow{NH_3} (EtGeN)_n$$

Primary and secondary amines that react only slowly with germanium halides combine more rapidly in the presence of a stronger tertiary base (Anderson, 1952a; Wieber and Frohning, 1967)—

$$\underset{\displaystyle NH_2}{\overset{\displaystyle NH_2}{\bigodot}} + Me_2GeCl_2 \xrightarrow{Et_3N} \underset{\displaystyle NH}{\overset{\displaystyle NH}{\bigodot}}GeMe_2 + 2Et_3NHCl$$

Other preparative methods that may be used in forming germanium—nitrogen bonds include the use of alkali metal or Grignard derivatives of amines (Baum et al., 1964; Satge and Baudet, 1966)—

$$Ph_3GeCl + Ph_2NLi \longrightarrow Ph_3GeNPh_2$$

$$Et_3GeCl + LiNH_2 \xrightarrow{THF} (Et_3Ge)_2NH$$

$$Et_3GeCl + PhN(MgBr)_2 \longrightarrow (Et_3Ge)_2NPh$$

This type of reaction applied to the dilithio-derivative of azobenzene and diphenyldichlorogermane produces a cyclic trimer, whereas a similar reaction between triphenylgermyl-lithium and azobenzene or azoxybenzene gives, after hydrolysis, the germylhydrazine (George $et\ al.$, 1960, 1966)—

$$\text{Ph}_2\text{GeCl}_2 + \text{Li}_2[\text{PhN=NPh}] \longrightarrow \text{Ph}_2\text{Ge} \underset{\text{N-N}}{\overset{\text{N-N}}{}} \text{GePh}_2$$

with Ph Ph on top (N—N) and Ph Ph on bottom (N—N)

$$\text{Ph}_3\text{GeLi} + \text{PhN=NPh} \longrightarrow \text{Ph}_3\text{GeN(Ph)N(Li)Ph}$$

$$\Big\downarrow \text{H}_2\text{O}$$

$$\text{Ph}_3\text{GeN(Ph)NHPh}$$

Diphenyldichlorosilane also yields a cyclic compound, but the corresponding reaction with diphenyldichlorostannane produces polymeric diphenyltin, formed by halogen–metal exchange and elimination of lithium halide—

$$\text{Ph}_2\text{SnCl}_2 + \text{Li}_2[\text{PhN=NPh}] \xrightarrow[\text{2. H}_2\text{O}]{\text{1. Et}_2\text{O}} (\text{Ph}_2\text{Sn})_n + \text{PhNHNHPh} + 2\text{LiCl}$$

Transammination reactions and elimination of water from amines and organogermanium oxides have also been used for preparative purposes—

$$(\text{Bu}^n_3\text{Ge})_2\text{NH} + \text{RNH}_2 \longrightarrow (\text{Bu}^n_3\text{Ge})_2\text{NR} + \text{NH}_3$$

$$(\text{R}_3\text{Ge})_2\text{O} + 2\text{HN}\square \longrightarrow 2\text{R}_3\text{Ge-N}\square + \text{H}_2\text{O}$$

Alkoxygermanes combine with phenylisocyanate in a mildly exothermic reaction with the formation of a germylcarbamate (Ishii $et\ al.$, 1967)—

$$\text{R}_3\text{GeOR}' + \text{PhNCO} \rightarrow \text{R}_3\text{GeN(Ph)CO}_2\text{R}'$$

Carbodi-imides react in the same way, but less readily—

$$\text{Bu}^n_3\text{GeOMe} + \text{RN=C=NR} \xrightarrow{100°\text{C}} \overset{\text{Bu}^n_3\text{Ge}}{\underset{\text{R}}{}}\!\!\!\!>\!\!\text{N-C}\!\!<\!\!\overset{\text{OMe}}{\underset{\text{NR}}{}}$$

A reaction that has not yet been applied to tetraphenylgermane is the ammonolysis by potassium amide in liquid ammonia; both tetraphenylsilane and -plumbane are completely degraded (Schmitz and Jansen, 1967)—

$$Ph_4Pb \xrightarrow{\text{KNH}_2,\ \text{NH}_3} K[Pb(NH_2)_3]$$

$$Ph_4Si \xrightarrow{\text{KNH}_2,\ \text{NH}_3} K_2 \begin{bmatrix} \begin{array}{c} (NH_2)_2 \\ Si-N \\ HN \diagdown \diagup Si(NH_2)_2 \\ Si-N \\ (NH_2)_2 \end{array} \end{bmatrix}^{2-}$$

More complex germanium–nitrogen compounds have been reported, such as $Me_3GeN(SiMe_3)_2$ (Scherer and Schmidt, 1964; Scherer and Biller, 1967) and $Ph_3Ge\overset{-}{N}\text{—}\overset{+}{P}Ph_3$ (Reichle, 1964).

Many tertiary germylamines are sufficiently stable to be distilled at temperatures up to 200°C. The reactions so far reported are concerned with the cleavage of the germanium—nitrogen bond by various protic reagents; all germanium–nitrogen compounds are susceptible to hydrolysis, although reactions are less rapid than with tin–nitrogen compounds. The reactivity towards cleavage by protic reagents decreases in the order $R_3GeNR'_2$ > $(R_3Ge)_2NH$ > $(R_3Ge)_3N$. Although no kinetic data are available, it is clear that hydrolysis rates depend markedly on the strength of the nitrogen base; the stronger the base, the greater the rate of hydrolysis (Rijkens et al., 1965). The most reasonable mechanism is rapid, and necessarily reversible, protonation at nitrogen, followed by separation of the amine as the rate-determining stage—

$$R_3GeNR_2 + H_2O \rightleftharpoons [R_3GeNHR_2]^+ + OH^-$$

$$[R_3GeNHR_2]^+ \rightarrow R_3Ge^+ + R_2NH$$

One of the most stable compounds is triphenylgermylpyrole, which may be isolated from aqueous solution. Germanium–nitrogen compounds also react with carboxylic acids, phenols, alcohols, mercaptans, and even with phenylacetylene, primary amines and diphenylphosphine under forcing conditions (Satge and Baudet, 1966). Butanethiol reacts exothermically with the amine, Me_3GeNEt_2 (Abel et al., 1966)—

$$Me_3GeNEt_2 + Bu^nSH \rightarrow Me_3GeSBu^n + Et_2NH$$

The addition of Me_3GeNR_2 to β-propiolactone has been examined kinetically (Itoh et al., 1967). The cyclic trimer $(Me_2GeNMe)_3$ reacts with methyllithium to give an N-lithiated monomer that has been used in the preparation of germanium–nitrogen–metal systems (Ruidisch and Mebert, 1968)—

$$(Me_2GeNMe)_3 \xrightarrow{\text{MeLi}} Me_3GeN(Li)Me \xrightarrow{\text{Me}_3\text{MCl}} Me_3GeN(Me)MMe_3$$

$$(M = C, Si, Ge, Sn, Pb)$$

4*

Reactions in which simple unsaturated molecules insert into the germanium–nitrogen bond have received some study (Satge *et al.*, 1964), and these parallel the behaviour of tin–nitrogen systems (Jones and Lappert, 1965)—

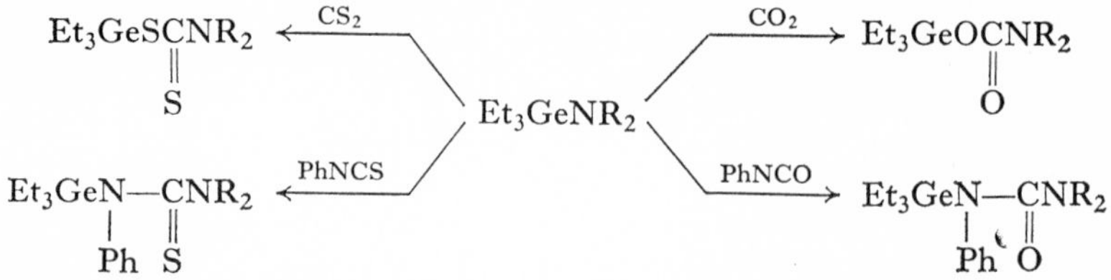
$Et_3GeSCNR_2$
S
CS_2
CO_2
$Et_3GeOCNR_2$
O
Et_3GeNR_2
$Et_3GeN—CNR_2$
Ph S
PhNCS
PhNCO
$Et_3GeN—CNR_2$
Ph O

Reactions of germanium–nitrogen compounds with metal hydrides are discussed in Chapter 8.

II. GERMANIUM–PHOSPHORUS COMPOUNDS

Sufficient is known of the chemistry of organogermanium–phosphorus complexes to indicate a general similarity to germanium–nitrogen compounds. For example, both types are readily hydrolysed, but phosphorus complexes are also highly sensitive to oxidation.

The first reported synthesis made use of the reaction (Brooks *et al.*, 1965)—

$$R_3GeX + Ph_2PLi \rightarrow R_3GePPh_2 + LiX$$

Lithium diphenylphosphide, which is obtained as a red solution in tetrahydrofuran by the action of lithium on diphenylphosphine, reacts exothermically with organogermanium monohalides to give the germylphosphine in high yield. Dihalides, R_2GeX_2, react in the same way, but with trihalides or germanium tetrachloride coloured polymeric material and tetraphenyldiphosphine are produced. These must result from halogen–metal exchange reactions, and possibly reduction of germanium(IV) to germanium(II). By contrast, both silicon and tin have yielded the tetrakis complexes $(Et_2P)_4Si$ and $(Ph_2P)_4Sn$. Secondary phosphines and phosphine, in the presence of a base, combine with germanium halides (Schumann *et al.*, 1966)—

$$Ph_3GeCl + PhPH_2 \xrightarrow{Et_3N} (Ph_3Ge)_2PPh$$

$$Ph_3GeCl + PH_3 \xrightarrow{Et_3N} (Ph_3Ge)_3P$$

A preliminary report shows that phosphine cleaves the germanium—nitrogen bond under mild conditions (Schumann and Blass, 1966). Diphenylphosphine is less reactive, but at 200°C it also cleaves germanium—nitrogen bonds—

$$Me_3GeNMe_2 + 3PH_3 \rightarrow (Me_3Ge)_3P + 3Me_2NH$$

Most of the information available on the reactivity of germylphosphines relates to the compound, Et_3GePPh_2. Its thermal stability is considerable and no decomposition is observed at 180°C in a nitrogen atmosphere. Hydrolysis by 10% aqueous monoglyme is complete within 5 min at 20°C, whereas the triphenylgermyl analogue, Ph_3GePPh_2 requires some 6 hr under similar conditions—

$$2Et_3GePPh_2 + H_2O \rightarrow (Et_3Ge)_2O + 2Ph_2PH$$

Oxidation by dry oxygen is a rapid reaction resulting in cleavage of the germanium—phosphorus bond giving a monomeric phosphorus(v) ester, which is only slowly hydrolysed by water—

$$Et_3GePPh_2 \xrightarrow{O_2} Et_3GeOPPh_2 \xrightarrow{H_2O} (Et_3Ge)_2O + Ph_2P(O)OH$$
$$\underset{\displaystyle O}{\overset{\displaystyle \|}{}}$$

Nucleophilic cleavage of the germanium—phosphorus bond by butyl-lithium is rapid at 0°C, as is its bromination—

$$Et_3GePPh_2 \begin{cases} \xrightarrow{Br_2} Et_3GeBr + Ph_2PBr \\ \xrightarrow{Bu^nLi} Et_3GeBu^n + Ph_2PLi \end{cases}$$

Methyl iodide cleaves the germanium—phosphorus bond, but in contrast to silicon and tin compounds, this reaction shows no sign of an intermediate phosphorus(v) salt analogous to $[Me_3SiPEt_3]I$. However the phosphorus atom does retain its donor character to an appreciable degree, since it forms a co-ordination complex with silver iodide—

$$Et_3GePPh_2 \begin{cases} \xrightarrow{MeI} Et_3GeI + [Ph_2PMe_2]^+I^- \\ \xrightarrow{AgI} [Et_3GePPh_2.AgI]_4 \end{cases}$$

Butyl-lithium also reacts with the tin—phosphorus bond and use has been made of this reaction to form a further phosphorus—metal bond—

$$(Ph_3Sn)_2PPh \xrightarrow{BuLi} [Ph_3SnPPh]Li \xrightarrow{Ph_3GeCl} Ph_3SnP(Ph)GePh_3$$

A further illustration of the donor character of the phosphorus atom in germylphosphines is the reaction with nickel carbonyl—

$$(Me_3Ge)_3P + Ni(CO)_4 \rightarrow (Me_3Ge)_3PNi(CO)_3 + CO$$

Silylphosphines behave in the same way producing complexes of lower thermal stability. The difference is ascribable to a higher degree of π-bonding in the silylphosphine making it a weaker electron donor (Schumann and Stelzer, 1967).

Further similarities to germanium—nitrogen compounds are shown by reactions with other protic reagents—

$$Et_3GePPh_2 \begin{cases} \xrightarrow[150°C]{NH_3} (Et_3Ge)_2NH + Ph_2PH \\ \xrightarrow{HCl} Et_3GeCl + Ph_2PH \end{cases}$$

Germanium–phosphorus compounds undergo insertion reactions with carbon disulphide, isocyanates and isothiocyanates, which are also analogous to those of nitrogen—

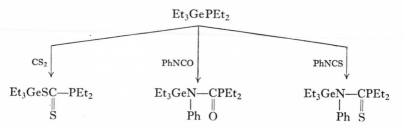

$$Et_3GePEt_2$$

| CS_2 | PhNCO | PhNCS |

$$Et_3GeSC—PEt_2 \quad\quad Et_3GeN—CPEt_2 \quad\quad Et_3GeN—CPEt_2$$
$$\overset{\|}{S} \quad\quad\quad\quad\quad \underset{Ph}{|} \ \overset{\|}{O} \quad\quad\quad\quad \underset{Ph}{|} \ \overset{\|}{S}$$

Carbon dioxide is unreactive even at 150°C under pressure. Acrylonitrile reacts exothermically with diethylphosphinotriethylgermane, and an i.r. examination of the product suggests that phosphorus is bonded to the terminal carbon atom—

$$Et_3GePEt_2 \xrightarrow{CH_2:CHCN} Et_3GeCH(CN)CH_2PEt_2$$

The reaction with phenylacetylene, initiated by azobisisobutyronitrile, produces both *cis–trans* addition and cleavage products (Satge and Couret, 1967)—

$$Et_3GePEt_2 + PhC\!:\!CH \xrightarrow{80°C} \begin{cases} Et_3GeC\!:\!CPh + Et_2PH \\ \\ Et_3GeCH\!:\!C(Ph)PEt_2 + Et_3GeC(Ph)\!:\!CHPEt_2 \end{cases}$$

$$\phantom{Et_3GePEt_2 + PhC\!:\!CH \xrightarrow{80°C}\quad\quad\quad} cis\text{–}trans \quad\quad\quad\quad\quad cis\text{–}trans$$

6 | ORGANOGERMANIUM–GROUP VI COMPOUNDS

I. GERMANIUM–OXYGEN COMPOUNDS

The reactivity of group IV metal–oxygen bonds increases progressively from silicon to lead. The differences are not very marked between silicon and germanium, even though the silicon—oxygen bond strength is considerably greater than that of germanium—oxygen. Some compounds, such as hexaphenyldigermoxane, even show a greater thermal stability than the corresponding disiloxane. Alkoxylead compounds are highly reactive, and even such weak acids as cyclopentadiene, chloroform and acetylene cleave the lead—oxygen bond. Tin—oxygen compounds also react with acetylene, but under more drastic conditions—

$$Et_3SnOMe + PhC\colon CH \xrightarrow{150°C} Et_3SnC\colon CPh + MeOH$$

The base strengths of $(R_3M)_2O$ compounds show considerable changes, and bistriethyltin oxide is a sufficiently strong base to adsorb carbon dioxide avidly, forming a crystalline carbonate—

$$(Et_3Sn)_2O + CO_2 \rightarrow (Et_3SnO)_2CO$$

Progressive trends are not always apparent; silanols are more stable than organotin hydroxides, but germanols are more readily dehydrated than either. Although hexaethyldistannoxane forms a low-melting solid on reaction with 1 mole of water, the hydrate is better formulated as a hydrated oxide $(Et_3Sn)_2O \cdot H_2O$ rather than the true hydroxide, Et_3SnOH (Lohmann, 1965).

A. Germanols

Relatively few germanols, R_3GeOH, are sufficiently stable to be isolated in the pure state, and evidence for the existence of germanediols, $R_2Ge(OH)_2$ has only recently become convincing. Rochow and Allred (1955) observed that dimethyldichlorogermane is completely hydrolysed by water, forming a clear acidic solution. On this evidence, the product could be the germanediol or the cyclic oxide $(Me_2GeO)_4$, which is also water soluble. The Raman spectrum of dimethylgermanium oxide in water provides evidence for the diol (Tobias and Hutcheson, 1966)—

$$(Me_2GeO)_4 + 4H_2O \rightleftharpoons 4Me_2Ge(OH)_2$$

101

Two intense lines are observed in the region of germanium–carbon stretching vibrations, one being polarized and the other depolarized. Hence the carbon–germanium–carbon skeleton is appreciably bent (effectively C_{2V} symmetry). Both the symmetric and antisymmetric Ge–(OH)$_2$ stretching vibrations are assigned to a line at 673 cm^{-1}, close to the assignment found in the complex ion $[O_2Ge(OH)_2]^{2-}$, which exists in alkaline solutions of germanium dioxide (Walrafen, 1965).

However octahedral complexes, such as the phthalocyanine derivative, in which hydroxyl groups are mutually *trans*, do exist and are only dehydrated with difficulty (Chapter 2). The instability of germanols is due to their ease of intermolecular dehydration, forming oxygen-bridged compounds of various types (digermoxanes, cyclic oligomers and polymers). Arylgermanols are more stable than alkyl-, and the only trialkylgermanols isolated are those with organic groups (e.g., isopropyl) of large steric requirement, and these owe their existence to the difficulty of bimolecular dehydration.

The only preparative method of importance is the hydrolysis of organogermanium halides which is usually carried out with caustic alkali in aqueous ethanol. Kinetic studies on this reaction are discussed in Chapter 7. Tri-1-naphthylbromogermane can be rapidly titrated with sodium hydroxide (West, 1952), and the resulting germanol may be converted into the digermoxane with Karl-Fischer reagent or by heating at 130°–140°C—

$$(1\text{-}C_{10}H_7)_3GeBr \xrightarrow{KOH} (1\text{-}C_{10}H_7)_3GeOH \longrightarrow [(1\text{-}C_{10}H_7)_3Ge]_2O$$

Triphenylgermanol, m.p. 133°C, is formed in the same way (Brook and Gilman, 1954). Tri-isopropylbromogermane is more resistant to hydrolysis and requires warming with 6M sodium hydroxide (Anderson, 1953).

Many other reactions produce germanols, at least as intermediates in hydrolysis reactions. Decarbonylation of germylcarboxylic acids produces either the germanol or an ester derived from it (p. 75). Oxidation of tricyclohexylgermane by passing air through a hot solution in carbon tetrachloride gives the corresponding germanol, which has been chlorinated by acetyl chloride (Johnson and Nebergall, 1949)—

$$R_3GeH \xrightarrow{O_2} R_3GeOH \xrightarrow{AcCl} (C_6H_{11})_3GeCl + AcOH$$

Although attempts to isolate trimethylgermanol have been unsuccessful (Schafer, 1961), the lithium salt has been obtained by the reactions (Ruidisch and Schmidt, 1963c)—

$$Me_3GeOGeMe_3 + MeLi \rightarrow Me_4Ge + Me_3GeOLi$$
$$(Me_2GeO)_4 + 4MeLi \rightarrow 4Me_3GeOLi$$

Above 100°C it decomposes to lithium oxide and the digermoxane—

$$2Me_3GeOLi \xrightarrow{>100°C} (Me_3Ge)_2O + Li_2O$$

The failure to obtain simple alkylgermanols leaves an unfortunate blank relative to the properties of silanols and stannols. For both of these the trialkyls are weaker acids and stronger bases than the corresponding aryls such as Ph_3SiOH and Ph_3SnOH; hence the net effect of aromatic rings is electron withdrawal.

Relative acid and base strengths of the triphenyl–Group IV metal hydroxides have been examined (West and Baney, 1960; West *et al.*, 1960; Matwiyoff and Drago, 1965). Acidities (proton-donating power) of Ph_3MOH compounds can be compared by observing the shift in oxygen—hydrogen stretching frequency on hydrogen bonding to a suitable base, such as ether. Relative basicities (proton-accepting power) of the oxygen atom in Ph_3MOH may be compared by adding a more powerful proton donor, such as phenol, and observing the change in the oxygen–hydrogen stretching frequency of phenol. The assumption is made that the magnitudes of the shifts $\Delta\nu(Ph_3MO—H)$ and $\Delta\nu(PhO—H)$ are proportional to the acidity and basicity respectively, of the triphenyl–metal hydroxide. Experimental values are shown in Table XIV. Two factors are involved in determining these relative acid and base strengths; the electronegativity of the

TABLE XIV

Hydrogen bonding in Ph_3MOH compounds

Ph_3MOH	$\nu(Ph_3MO—H)$ cm^{-1}	Relative basicity, $\Delta\nu(O—H)$ cm^{-1} PhOH	Relative acidity, $\Delta\nu(Ph_3MO—H)$ cm^{-1} Et$_2$O
Ph_3COH	3609	172	174
Ph_3SiOH	3677	175	316
Ph_3GeOH	3651	288	198
Ph_3SnOH	3647	470	$\langle 10$

group IV element bonded to oxygen and the extent of d_π–p_π bonding between doubly occupied p orbitals of oxygen and unoccupied d orbitals of silicon, germanium or tin. The large increase in acidity between Ph_3COH and Ph_3SiOH is a reflection of extensive π bonding in the silanol, since the decrease in electronegativity between carbon and silicon would have the reverse effect. Triphenylgermanol is slightly more acidic than triphenylmethanol, which points to π-bonding between germanium and oxygen more than offsetting the difference expected on the basis of their electronegativities.

The difference between the acid strengths of the silanol and germanol is probably a result of poorer overlap and a greater energy difference between the germanium $4d$ and oxygen $2d$ orbitals relative to the silicon $3d$ orbitals. The basicities (C $\sim$ Si $<$ Ge $<$ Sn) are also understandable in terms of relative electronegativities and π-bonding characteristics of the Group IV element. In the absence of π-bonding, the greater the electronegativity the weaker the base strength. Again, if the electronegativities of silicon and germanium are considered as essentially the same, then the difference in basicity between Ph_3SiOH and Ph_3GeOH reflects less extensive π-bonding in the germanol.

Dehydration of germanols to digermoxanes proceeds readily. Even tri-isopropylgermanol, which is sterically hindered, has been dehydrated at 200°C, and triphenylgermanol dehydrates in boiling toluene.

B. Digermoxanes

Digermoxanes are formed by the hydrolysis of organogermanium mono-halides via the germanol—

$$R_3GeX \xrightarrow{H_2O} R_3GeOH \longrightarrow (R_3Ge)_2O$$

(Anderson, 1951a, b; Gilman and Leeper, 1951; Fuchs and Gilman, 1958; Emel'yanova et al., 1962; Rijkens and Van der Kerk, 1964). The usual procedure is to shake or heat the halogermane with 10–20% aqueous potassium hydroxide, ammonia or dry silver carbonate, and then extract the digermoxane into an organic solvent. Germanium hydrido chlorides are also hydrolysed in this way, giving compounds such as $[Et(H)_2Ge]_2O$ (Massol and Satge, 1966). Many other substituted germanes may be hydrolysed to digermoxanes; these include pseudohalides, nitrogen, phosphorus and transition-metal derivatives. Decarbonylation of esters such as $Ph_3GeOCOGePh_3$ and oxidation of germanium–mercury and related compounds produce digermoxanes. Tetraethyldichlorodigermane, $Et_4Ge_2Cl_2$, formed in the redistribution reaction between hexaethyldigermane and germanium tetrachloride has been characterized as the 1,2-dichloro-isomer by its hydrolysis to a cyclic digermoxane (Bulten and Noltes, 1966a)—

$$Et_2(Cl)GeGe(Cl)Et_2 \xrightarrow{NaOH} \begin{array}{c} Et_2\ \ Et_2 \\ Ge{-}Ge \\ O \qquad\quad O \\ Ge{-}Ge \\ Et_2\ \ Et_2 \end{array}$$

Cleavage of germanium—carbon bonds by electrophiles commonly yields digermoxanes after hydrolysis. For example, a methyl group can be cleaved from trimethyl(chloromethyl)germane by concentrated sulphuric acid, giving a sulphonic ester from which the digermoxane is obtained by hydrolysis (Mironov *et al.*, 1964b)—

$$Me_3GeCH_2Cl \xrightarrow[\text{2. } H_2O]{\text{1. } H_2SO_4 \text{ (110°C)}} [Me_2(ClCH_2)Ge]_2O$$

Kinetic studies on the acid and base solvolysis of arylgermanium bonds are discussed in Chapter 3.

Digermoxanes are of high thermal stability; several have been distilled at temperatures above 300°C without decomposition. The volatile digermoxanes have a strong camphor-like smell. Few chemical reactions have been studied other than their conversion into halides and their reduction by lithium aluminium hydride. Hexamethyldigermoxane is cleaved by boron trifluoride (Griffiths and Onyszchuk, 1961)—

$$(Me_3Ge)_2O \xrightarrow{BF_3} 2Me_3GeF + B_2O_3$$

Low-temperature bromination of germanium—hydrogen bonds in $[Et_2(H)Ge]_2O$ can be achieved without breaking the digermoxane structure, and oxidation gives diethylgermanium oxide—

$$[Et_2(H)Ge]_2O \begin{cases} \xrightarrow[0°C]{Br_2, \ EtBr} [Et_2(Br)Ge]_2O \\ \xrightarrow{O_2} (Et_2GeO)n \end{cases}$$

C. Diorganogermanium oxides

The hydrolysis of organogermanium dihalides produces oxides having the composition R_2GeO that exist as cyclic trimers, tetramers and polymers. Alkyldichlorogermanes, $RGe(H)Cl_2$, only yield polymers (Anderson, 1961; Emel'yanova *et al.*, 1962; Massol and Satge, 1966)—

$$Bu^n_2GeBr_2 \xrightarrow{NaOH} \begin{array}{c} Bu^n_2 \\ O-Ge \\ Bu^n_2Ge \qquad O \\ O-Ge \\ Bu^n_2 \end{array}$$

$$Et_2GeBr_2 \xrightarrow{NaOH} \begin{array}{c} Et_2 \\ Ge-O \\ O \qquad GeEt_2 \\ Et_2Ge \qquad O \\ O-Ge \\ Et_2 \end{array}$$

The first study of the relationship and interconversion of these oxides was made by Metlesics and Zeiss (1960), who obtained an insoluble amorphous diphenylgermanium oxide by the hydrolysis of diphenyldibromogermane with silver nitrate. If this crude polymer is boiled with acetic acid, it yields a cyclic tetramer that, on distillation, re-arranges to a cyclic trimer. The cyclic tetramer is converted into a presumably linear polymer in ethanol–water and this process also may be reversed—

$$(Ph_2GeO)_n \underset{EtOH-H_2O}{\overset{HOAc}{\rightleftarrows}} (Ph_2GeO)_4 \underset{HOAc}{\overset{dist}{\rightleftarrows}} (Ph_2GeO)_3$$

m.p. 298°C   m.p. 218°C   m.p. 149°C

Rearrangement reactions of this type which involve cleavage of germanium–oxygen bonds have been more closely studied for dimethylgermanium oxide. This also has been obtained as a trimer, tetramer and polymer. Tensimetric measurements show an average degree of polymerization of 3 at high temperature (250°C) and low pressure (100 mm), which rises to 3·2 at lower temperature (160°C) and higher pressure (Brown and Rochow, 1960). In the solid state, the crystalline trimer and tetramer are slowly converted into the polymeric form, but depolymerization takes place on heating with a solvent. P.m.r. studies show that the only forms present in solution are the cyclic trimer and tetramer, and in inert solvents these slowly reach equilibrium (Moedritzer, 1966b)—

$$4(Me_2GeO)_3 \rightleftharpoons 3(Me_2GeO)_4$$

High concentration and low temperature shift the equilibrium towards the tetramer, whereas high dilution and high temperature favour trimer formation. The enthalpy of formation of the tetramer has been evaluated as −8·2 kcal. mole^{-1}.

Few reactions have been reported on diorganogermanium oxides other than their conversion into dihalides using mineral acids, and their reaction with methyl-lithium. Trimeric di-n-butylgermanium oxide decomposes when boiled at 760 mm (Anderson, 1961). Tetrameric dimethylgermanium oxide and dimethylgermanium dihalides equilibrate slowly, and at equilibrium, cyclic and chain molecules co-exist. The number of species present is very large, but one type may be represented by (Moedritzer and Van Wazer, 1965b)—

$$(Me_2GeO)_4 + Me_2GeCl_2 \rightleftharpoons Me_2(Cl)Ge \left(O \underset{Me_2}{\overset{}{Ge}} O \right)_3 Ge(Cl)Me_2$$

D. Germanoic anhydrides

Hydrolysis of organogermanium trihalides either directly or by ammonolysis followed by hydrolysis produces polymeric germanoic anhydrides

(Flood, 1933; Anderson, 1950; Johnson and Jones, 1952; Emel'yanova *et al.*, 1962). Organogermanium trihydrides are hydrolysed by strong base in the same way (Satge, 1961)—

$$(cyclohexyl)GeH_3 \xrightarrow{\text{KOH, EtOH}} [(cyclohexyl)GeO)_2O]_n + H_2$$

The parent acids, $RGe(O)OH$, are unknown although an elemental analysis has been reported corresponding to $m\text{-}MeC_6H_4Ge(O)OH$, but without further evidence of structure. Little is known of the properties of germanoic anhydrides; they decompose without melting at high temperatures ($>300°C$) and are therefore polymeric. *In vacuo* above $300°C$, ethylgermanoic anhydride decomposes with the formation of an unidentified sublimate. When an alkyltrihalogermane is added to water a white oily film is formed that dissolves on standing leaving a clear acid solution from which the germanoic anhydride separates as a Cellophane-like film. When dry, they form white amorphous powders, soluble in organic solvents, such as ethanol and carbon tetrachloride, and are freely soluble in sodium hydroxide solution. The only reaction reported is their quantitative conversion into trihalides by treatment with halogen acids in strong aqueous solution.

E. Germoxy–metal compounds

Various compounds have been described that contain the grouping Ge–O–M; their properties, such as the basicity of the oxygen atom relative to digermoxanes, have not been reported, but several cleavage reactions have been studied, and these reflect the relative metal—oxygen bond strengths.

Germoxy-silanes and -stannanes have been prepared by two methods: (Seyferth and Alleston, 1963; Schmidbaur and Hussek, 1964)—

$$Me_3GeOLi + Me_3SiCl$$
$$Me_3GeOSiMe_3$$
$$Me_3GeCl + Me_3SiOLi$$

Dimethyldichlorogermane has yielded both mono- and di-siloxy derivatives. Germoxy-aluminium, -gallium and -indium compounds are dimeric in solution. This association, which is typical of the electron-deficient Group III metal alkyls, produces 4-membered ring systems in which the Group III metal is tetra-co-ordinated (Armer and Schmidbaur, 1967)—

$$2R_3GeOLi + 2R'_2GaCl \longrightarrow R_3Ge-O \underset{\underset{R'_2}{Ga}}{\overset{\overset{R'_2}{Ga}}{\diamond}} O-GeR_3$$

Germoxysilanes are rapidly hydrolysed by water, acids and bases, but an exception is the tetrakis compound $Ge(OSiPh_3)_4$, which is stable (as the crystalline solid) to concentrated hydrochloric acid and sodium hydroxide solution for short periods. The symmetry of this molecule is reflected in its high melting point of 472°C (Gutmann and Meller, 1960). Germoxy-silanes slowly disproportionate at their boiling points—

$$2R_3GeOSiR_3' \rightarrow (R_3Ge)_2O + (R_3'Si)_2O$$

The tin compound $Me_3GeOSnMe_3$ disproportionates more readily, and both types are highly toxic. Aluminium chloride and phenyl-lithium both cleave the weaker germanium—oxygen bond in the germoxysilanes (Schmidbaur and Schmidt, 1961)—

$$Me_3GeOSiMe_3 \begin{cases} \xrightarrow{H_2O} (Me_3Ge)_2O + (Me_3Si)_2O \\ \xleftarrow{AlCl_3} Me_3GeCl + Me_3SiOAlCl_2 \\ \xrightarrow{PhLi} Me_3GePh + Me_3SiOLi \end{cases}$$

With sulphur trioxide and chromium trioxide they form esters, Me_3-$GeOSO_2OSiMe_3$ and $Me_3GeOCrO_2OSiMe_3$. Although aluminium chloride cleaves the Ge–O–Si group, it has been used to co-condense triphenyl-germanol and -silanol, producing a waxy solid by cleavage of one phenyl—silicon bond with subsequent dehydration reactions (Delman *et al.*, 1966)—

$$2Ph_3GeOH + 2Ph_3SiOH \xrightarrow{AlCl_3} Ph_3Ge \overset{O}{\diagup} \underset{Ph_2}{Si} \overset{O}{\diagup} \underset{Ph_2}{Si} \overset{O}{\diagup} GePh_3$$

F. Alkoxides

Many alkoxygermanes have been reported, ranging from the type $R_nGe(OR)_{4-n}$ to heterocyclic compounds in which the Ge–O–C grouping forms part of a 5- or 6-membered ring. Most alkoxides are liquids and are readily hydrolysed. The donor character of the oxygen atom has been demonstrated for trimethylmethoxygermane, which forms a crystalline 1:1 adduct with boron trifluoride.

Sodium alkoxides convert organogermanium halides into the corresponding alkoxides (West *et al.*, 1954; Griffiths and Onyszchuk, 1961)—

$$R_nGeCl_{4-n} + R'ONa \xrightarrow{R'OH} R_nGe(OR')_{4-n} + NaCl$$

A more satisfactory procedure is the reaction between a germanium-halide, -oxide or germanoic anhydride and the alcohol in the presence of a base, such as triethylamine. Water formed in the latter reactions may be separated by azotropic distillation with benzene (Wieber and Schmidt, 1963a, b; Mehrotra and Mathur, 1966, 1967; Wieber and Frohning, 1967)

$$RGeCl_3 + 3R'OH \xrightarrow{NH_3} RGe(OR')_3 + 3NH_4Cl$$

$$[(BuGeO)_2O]_n + 3R'OH \xrightarrow{C_6H_6} BuGe(OR')_3 + 3H_2O$$

This type of reaction has been applied to the formation of heterocyclic compounds—

$$(R_2GeO)_3 + HO(CH_2)_2NH_2 \xrightarrow{C_6H_6} \quad + H_2O$$

Acetylacetone reacts with dibutylgermanium oxide to yield an orange–red liquid, which may be a hexaco-ordinate complex—

$$(Bu^n_2GeO)_3 + MeCOCH_2COMe \longrightarrow$$

Alkoxide exchange reactions are readily applied when the displaced alcohol has the greater volatility (Mehrotra and Mathur, 1967)—

$$BuGe(OEt)_3 + 3ROH \rightarrow BuGe(OR)_3 + 3EtOH$$

Epoxides are cleaved by organogermanium halides (Lavigne et al., 1967)—

$$Pr^n_3GeBr + Me \longrightarrow Pr^n_3GeOCH_2CH(Br)Me$$

The decarbonylation of organogermanium carboxylic esters is of interest mechanistically (Brook, 1955; Brook and Peddle, 1963)—

$$R_3GeOR + CO$$

This reaction as applied to an asymmetric germane, occurs with retention of configuration, which is consistent with a unimolecular 1,2-rearrangement.

The reaction of organogermanium hydrides with alcohols is catalysed by copper powder, in the same way as their reaction with water (Lesbre and Satge, 1962; Satge, 1964). Phenols eliminate hydrogen more readily than alcohols, and with dihydrides the reaction can be controlled to produce alkoxy hydrides—

$$Bu^n_2GeH_2 + ROH \xrightarrow[100°]{Cu} Bu^n_2Ge(H)OR + H_2$$

$$Et_3GeH + PhOH \xrightarrow[80°]{Cu} Et_3GeOPh + H_2$$

These reactions probably proceed by an ionic mechanism involving attack by oxygen on germanium—

$$\begin{array}{c} R_3Ge—H \\ \vdots \\ O—H \\ / \\ R' \end{array} \xrightarrow{Cu} R_3GeOR' + H_2$$

The catalytic activity of copper has not been explained, but it shows considerable specificity in reactions with unsaturated alcohols where other metals catalyse the addition of GeH to the multiple bond—

$$R_3GeH + CH_2:CHCH_2OH \begin{cases} \xrightarrow{H_2PtCl_6} R_3Ge(CH_2)_3OH \\ \xrightarrow{Cu} R_3GeOCH_2CH:CH_2 + H_2 \end{cases}$$

$$Et_3GeH + \underset{O \quad CH_2OH}{\bigcirc} \xrightarrow{Cu} Et_3GeOCH_2 \underset{O}{\bigcirc} + H_2$$

Germanium hydrides add to the carbonyl group of aldehydes either in the presence of copper powder or platinum

$$Et_3GeH + PhCHO \xrightarrow{Cu} Et_3GeOCH_2Ph$$

Exchange reactions between methoxy and other anionic ligands, such as chloride in substituted germanes are rapid and, for dimethyldimethoxygermane, the equilibrium concentration of the mixed product is large (Moedritzer and Van Wazer, 1965b)—

$$Me_2Ge(OMe)_2 + Me_2GeX_2 \rightleftharpoons 2Me_2Ge(Cl)OMe$$

G. Esters of carboxylic acids

A large number and variety of organogermanium esters have been described. Most preparative methods are of wide application, and the choice depends primarily on the available starting materials. Germanium tetra-chloride does not react with hot acetic acid, but several organogermanium halides form esters with free carboxylic acids in the presence of a base. Peroxyacids also produce esters with organogermanium halides, presumably via an unstable peroxy intermediate (Davis and Hall, 1959)

$$Pr^n_3GeCl + C_7H_{15}CO_3H \rightarrow Pr^n_2Ge\overset{\displaystyle Pr^n}{\underset{\displaystyle OCOC_7H_{15}}{\vert\vert}}O \rightarrow Pr^n_2Ge(OPr^n)OCOC_7H_{15}$$

Silver and lead salts of carboxylic acids have been more widely used in reactions with germanium halides. Whereas germanoic anhydrides do not react with acid anhydrides, triesters may be prepared via silver or lead salts (Anderson, 1952b, 1956b)—

$$EtGeI_3 \left\langle \begin{array}{l} \xrightarrow{Pb(OCHO)_2} EtGe(OCHO)_3 \\ \xrightarrow{AgOAc} EtGe(OAc)_3 \end{array}\right.$$

Another general method for preparing mono- or di-esters is the reaction between the free acid or acid anhydride and an organogermanium-oxide or -hydroxide. The stronger the acid the more rapid the reaction, and di-germoxanes react more rapidly than the less basic cyclic oxides, $(R_2GeO)_n$ (Anderson, 1950, 1951b). For example, formic acid and trifluoroacetic acid combine exothermically with hexaethyldigermoxane, but less readily with tetrameric diethylgermanium oxide—

$$(Et_3Ge)_2O \xrightarrow{HCO_2H} 2Et_3GeOCHO + 2H_2O$$

$$(Et_2GeO)_4 \xrightarrow{HCO_2H} 4Et_2Ge(OCHO)_2 + 4H_2O$$

Tricyclohexylgermanol is converted into the acetate by heating with acetic anhydride (Johnson and Nebergall, 1949).

Organogermanium alkoxides are readily converted into esters by heating with an organic acid. This reaction can be controlled so as to produce alkoxy esters in high yield (Mathur and Mehrotra, 1967)—

$$R_3GeOR + R'CO_2H \rightarrow R_3GeOCOR' + ROH$$

$$R_2Ge(OR)_2 + R'CO_2H \rightarrow R_3Ge(OR)OCOR' + ROH$$

Triethylgermane does not react with acetic acid alone, but in the presence of copper powder hydrogen is evolved with formation of the ester (Anderson, 1957; Lesbre and Satge, 1962). However stronger acids, such as trifluoro-acetic, do react—

$$Et_3GeH + F_3CCO_2H \rightarrow Et_3GeOCOCF_3 + H_2$$

Silver acetate is reduced by di-isopropylgermane—

$$Pr^i_2GeH_2 \xrightarrow{AgOAc} Pr^i_2Ge(OAc)_2 + Ag + H_2$$

Triethylgermanium acetate is the probable product when m- or p-(bistriethylgermyl)benzene is heated with a mixture of nitric acid, acetic acid and acetic anhydride (Eaborn et al., 1960)—

In trans-esterification reactions, although the equilibrium favours the ester derived from the stronger acid, they can be carried out so that the more volatile acid is displaced (Anderson, 1952b, 1955)—

$$R_3GeOCOR' + R''CO_2H \rightleftharpoons R_3GeOCOR'' + R'CO_2H$$

Most tetraorganogermanes are unreactive towards carboxylic acids, but germanacyclanes react readily with mono- or di-chloroacetic acids (Mazerolles et al., 1966)—

Similarly organogermanes containing unsaturated groups such as vinyl, allyl and ethynyl, are readily cleaved—

$$Bu^n_3GeCH_2C\vdots CH \xrightarrow{R'CO_2H} Bu^n_3GeOCOR' + MeC\vdots CH$$

The acid solvolysis of arylgermanium bonds on which there have been several kinetic studies are discussed in Chapter 3.

Esters formed from organic acids are of high thermal stability, and all types are quite readily hydrolysed. Some qualitative observations have been reported by Anderson (1950) who found that in the hydrolysis of the diacetate and diformate $[R_2Ge(OCOR)_2]$ 30% acetic acid produces a homogeneous solution whereas at the same concentration a 70% solution of formic acid is required.

A kinetic study of the hydrolysis of triorganogermanium acetates in aqueous dioxan shows that they are more rapidly hydrolysed than silicon analogues, possibly by a factor of about 10^3 and that the reactions are strongly catalysed by potassium acetate (Prince and Tims, 1967). It has been established by using isotopically labelled water ($H_2^{18}O$) that the metal—oxygen bond is cleaved during the hydrolysis. Relative rate constants vary according to the R groups bonded to germanium. At 25°C using 5% water in dioxan values are: Pr^n, 760; $n-C_6H_{13}$, 450; Ph, 260; Pr^i 7·3; C_6H_{11}, 1.

These variations are the net result of changes in π-bonding, inductive effects, steric hindrance and mass factors. The kinetics are compatible with rearward attack by water polymer at the germanium atom—

$$
\begin{array}{c}
\text{H} \\
\diagdown \\
\hspace{2em}\text{O} \text{---} \text{Ge} \text{---} \text{OAc} \\
\diagup \hspace{2.5em} | \\
\text{H} \\
\vdots \\
(\text{H}_2\text{O})_n
\end{array}
$$

Few other chemical reactions have been reported on organogermanium esters. Chlorinative cleavage of the germanium–oxygen bond has been observed—

$$\text{EtGe(OAc)}_3 + \text{Ph}_2\text{SiCl}_2 \rightarrow \text{EtGeCl}_3 + \text{Ph}_2\text{Si(OAc)}_2$$

H. Esters with inorganic acids

Strong inorganic acids form thermally stable esters with organogermanes which are all essentially covalent compounds. Organogermanium-oxides, -halides and -hydrides form esters on treatment with inorganic acid, their silver salts, or in some cases their anhydrides—

$$(\text{R}_3\text{Ge})_2\text{O} + \text{H}_2\text{SO}_4 \rightarrow (\text{R}_3\text{GeO})_2\text{SO}_2 + \text{H}_2\text{O}$$
$$\text{R}_3\text{GeCl} + \text{AgNO}_3 \rightarrow \text{R}_3\text{GeONO}_2$$
$$(\text{R}_3\text{Ge})_2\text{O} + \text{Cl}_2\text{O}_7 \rightarrow \text{R}_3\text{GeOClO}_3$$

Other less general methods are referred to under the type of ester.

Whereas digermoxanes, such as $(\text{Et}_3\text{Ge})_2\text{O}$, react with concentrated sulphuric acid on warming, cyclic oxides of the type $(\text{Et}_2\text{GeO})_4$ are more vigorously esterified (Anderson, 1950)—

$$(\text{Et}_3\text{Ge})_2\text{O} + \text{H}_2\text{SO}_4 \longrightarrow (\text{Et}_3\text{GeO})_2\text{SO}_2 + \text{H}_2\text{O}$$

$$(\text{Et}_2\text{GeO})_4 \xrightarrow{\text{H}_2\text{SO}_4} \text{Et}_2\text{Ge} \begin{array}{c} \diagup \text{O---SO}_2\text{---O} \diagdown \\ \diagdown \text{O---SO}_2\text{---O} \diagup \end{array} \text{GeEt}_2$$

Although this direct esterification is a general reaction, trans-esterification procedures using an ester of an organic acid give cleaner products in high yield (Anderson, 1951b, 1952b, 1955, 1964)—

$$\text{Pr}^n_2\text{Ge(OAc)}_2 \xrightarrow{\text{H}_2\text{SO}_4} [(\text{Pr}^n_2\text{GeO})_2\text{SO}_2]_2 + \text{HOAc}$$
$$\text{Pr}^n_3\text{GeOCOR} \xrightarrow{\text{H}_2\text{SO}_4} (\text{Pr}^n_3\text{GeO})_2\text{SO}_2 + \text{RCO}_2\text{H}$$

Triphenylgermane combines with benzene sulphonic acid to yield the corresponding ester—

$$\text{Ph}_3\text{GeH} \xrightarrow{\text{PhSO}_3\text{H}} \text{Ph}_3\text{GeOSO}_2\text{Ph} + \text{H}_2$$

Carbon—germanium bonds in heterocyclic germanium compounds have been cleaved by concentrated sulphuric acid (Mazerolles, 1962)—

$$R_2Ge\text{<ring>} \xrightarrow{H_2SO_4} R_2Ge\begin{smallmatrix}\nearrow Bu^n \\ \searrow OSO_2OH\end{smallmatrix}$$

Treatment of the germoxysilane $Me_3GeOSiMe_3$ with sulphur trioxide produces a mixed sulphonic ester (Schmidbaur and Schmidt, 1961)—

$$Me_3GeOSiMe_3 \xrightarrow{SO_3} Me_3GeOSO_2OSiMe_3$$

Electrophilic cleavage of arylgermanium bonds by sulphur trioxide also gives a sulphonic ester. In the following example, the presence of one ester group deactivates the molecule and the remaining aryl—germanium bond is unreactive towards sulphur trioxide (Bott et al., 1963b)—

$$Et_3Ge\text{—}\langle C_6H_4\rangle\text{—}GeEt_3 \xrightarrow{SO_3} Et_3Ge\text{—}\langle C_6H_4\rangle\text{—}SO_2OGeEt_3$$

All types of sulphonic esters are of high thermal stability; some can be distilled at temperatures as high as 380°C. They are covalent, and the ester $Bu^n_3GeOSO_2Et$ in acetic acid has a molar conductance of only 8×10^{-3} $ohm^{-1}\,mole^{-1}$. Sulphonic esters are readily hydrolysed, but the only equilibrium data relates to $Bu^n_3GeOSO_2Et$, which as a $0\cdot03$ M solution in 59% aqueous ethanol has a pH of $1\cdot9$.

Various other esters have been described, all of which are readily hydrolysed by water. Trimethylchlorogermane and silver nitrate react in dry tetrahydrofuran to give the nitrate ester as a liquid that can be distilled in vacuo (Schmidt and Ruidisch, 1962). In the same way, silver phosphate or phosphorus pentoxide have been used to prepare phosphate esters—

$$Me_3GeCl + AgNO_3 \xrightarrow{THF} Me_3GeONO_2 + AgCl$$

$$(Me_3Ge)_2O \xrightarrow{Ag_3PO_4} (Me_3GeO)_3PO \xleftarrow{P_2O_5} Me_3GeCl$$

When strongly heated, the phosphate decomposes to a polymeric metaphosphate, $(Me_3GeOPO_2)_n$. Similar reactions have been used to form arsenates, selenates, chlorophosphates, pervanadates, perrhenates and perchlorates (Schmidt et al., 1961a, b; Ruidisch and Schmidt, 1963d; Srivastava and Tandon, 1967)—

$$(Me_3Ge)_2O \begin{cases} \xrightarrow{As_2O_5} (Me_3GeO)_3AsO \\ \xrightarrow{SeO_3} (Me_3GeO)_2SeO_2 \\ \xrightarrow{Re_2O_7} (Me_3GeO)_2ReO_2 \\ \xrightarrow{Cl_2O_7} Me_3GeOClO_3 \end{cases}$$

Reactions with selenium trioxide and dichlorine heptoxide can produce explosions, and these esters are more safely prepared from the silver salts of the oxyacids. The germanium—oxygen bond in $Me_3GeOSiMe_3$ is mainly cleaved by phosphorus oxychloride—

$$Me_3GeOSiMe_3 + POCl_3 \begin{cases} \nearrow Me_3GeCl + Me_3SiOPOCl_2 \quad (80\%) \\ \searrow Me_3SiCl + Me_3GeOPOCl_2 \quad (20\%) \end{cases}$$

Trialkylgermylphosphinites have been briefly reported (Issleib and Walther, 1967)—

$$R_2POK + R_3GeCl \rightarrow R_3GeOPR_2 + KCl$$

I. Peroxy compounds

Organogermanium peroxides, which resemble silicon analogues, were first prepared by Davis and Hall (1959), who found that alkyl hydroperoxides and organochlorogermanes interact in the presence of an anhydrous base, such as ammonia or triethylamine—

$$Pr^n_3GeCl + Me_3COOH \xrightarrow{NH_3} Pr^n_3GeOOCMe_3 + NH_4Cl$$

Bis-germylperoxides and germyl hydroperoxides have similarly been obtained with hydrogen peroxide (Rieche and Dahlmann, 1964; Dannley and Farrant, 1966)—

$$Ph_3GeBr \begin{cases} \xrightarrow{H_2O_2,\ NH_3} Ph_3GeOOGePh_3 + NH_4Br \\ \xrightarrow{H_2O_2,\ NH_3} Ph_3GeOOH + NH_4Br \end{cases}$$

Other derivatives of organogermanes react with peroxydic compounds; these include alkoxides, oxides and hydroxides—

$$R_3GeOR' + RO_2H \rightleftharpoons R_3GeOOR + R'OH$$

$$Ph_3GeOH[\text{or } (Ph_3Ge)_2O] \xrightarrow[NH_3]{98\% \ H_2O_2} Ph_3GeOOH$$

Alkali-metal salts of peroxides have also been used—

$$R_3GeX + R'O_2Na \rightarrow R_3GeOOR' + NaCl$$

One example of a tetrakisperoxygermane has been obtained by the reaction between germanium tetrachloride and decahydronaphthylhydroperoxide—

$$GeCl_4 + C_{10}H_{17}OOH \xrightarrow{Et_3N} Ge(OOC_{10}H_{17})_4$$

It forms a white crystalline solid melting without decomposition at 85°C.

Many peroxides of the type R_3GeOOR' may be vacuum distilled and are stable up to about 70°C for short periods. Triphenylgermyl hydroperoxide is even more stable and shows only slight decomposition over 30 days at 25°C. All types of peroxygermanes are readily hydrolysed by water—

$$R_3GeO_2R' + H_2O \rightleftharpoons (R_3Ge)_2O + R'O_2H$$

They catalyse the polymerization of vinyl monomers, indicating that homolysis of the oxygen—oxygen bond can occur. Anhydrous hydrogen chloride cleaves the germanium—oxygen bond—

$$R_3GeO_2R' + HCl \rightarrow R_3GeCl + R'O_2H$$

None of these studies refers to any explosive hazard in handling peroxygermanes.

II. GERMANIUM–SULPHUR, –SELENIUM AND –TELLURIUM COMPOUNDS

Germanium–sulphur compounds of the type $Ge(SR)_4$ have already been considered in Chapter 2, and this discussion is restricted to compounds having at least one germanium—carbon bond (Abel and Armitage, 1967). All of the group IV elements form various compounds with sulphur, and, apart from silicon, they are often stable to water and may even be formed in aqueous solution. The majority of germanium–sulphur compounds are evidently hydrolysed in acid solution, although the amount of information available is limited. There is no example of a germanium—sulphur double bond; compounds in which this might be formed, such as R_2GeS, always associate, so that the germanium remains tetraco-ordinate.

The types of organogermanium sulphur compounds reported include R_3GeSH, $R_nGe(SR')_{4-n}$ $(R_3Ge)_2S$, $(R_3GeS)_2$, $(R_2GeS)_n$, $R_2Ge(X)SR'$. There are also various compounds in which germanium and sulphur form part of heterocyclic systems.

Germanium thiols have been prepared by the action of hydrogen sulphide on the metal halide in the presence of a base such as pyridine (Davidsohn et al., 1965)—

$$R_3GeX + H_2S \xrightarrow{py} R_3GeSH + pyHX$$

The germanium-thiol Ph_3GeSH is a white crystalline solid stable in dry air, but hydrolysed by water. These compounds may be obtained by the action of sulphur on a germanium hydride (Vyazankin *et al.*, 1966d)—

$$Et_3GeH + S \xrightarrow{140°C} Et_3GeSH$$

Alkali-metal salts of germane thiols are most readily obtained by the reaction (Henry and Davidsohn, 1962, 1963; Ruidisch and Schmidt, 1963b; Schumann *et al.*, 1965)—

$$Ph_3GeLi + S \rightarrow Ph_3GeSLi$$

A further elegant method, which has been applied to Me_3GeSLi is the action of methyl-lithium on the cyclic trimer $(Me_2GeS)_3$—

$$(Me_2GeS)_3 + 3MeLi \rightarrow 3Me_3GeSLi$$

Few reactions have so far been reported on the parent thiols, though they may be used to make "mixed" metal sulphides—

$$Et_3GeSH + Et_3SnH \rightarrow Et_3GeSSnEt_3 + H_2$$

Oxidation and alkylation reactions have been reported—

$$Ph_3GeSH \underset{RX, \, py}{\overset{I_2}{\underset{\nearrow}{\searrow}}} \begin{matrix} (Ph_3GeS)_2 + 2HI \\ \\ Ph_3GeSR + HX \end{matrix}$$

The lithio-derivative, Me_3GeSLi decomposes above 85°C, and has been used as a route to the "mixed" metal sulphides—

$$2Me_3GeSLi \xrightarrow{85°C} (Me_3Ge)_2S + Li_2S$$

$$Me_3GeSLi + Me_3SiCl \rightarrow Me_3GeSSiMe_3$$

These germanium–sulphur–silicon and germanium–sulphur–tin compounds disproportionate readily on heating—

$$2Me_3GeSSiMe_3 \rightarrow (Me_3Ge)_2S + (Me_3Si)_2S$$

The isolation of Ph_3GeSNa as a hygroscopic solid, soluble in benzene, alcohol and water has been achieved with hydrated sodium sulphide—

$$Ph_3GeBr + Na_2S \xrightarrow{EtOH} Ph_3GeSNa$$

The conditions are clearly somewhat critical (Hooton and Allred, 1965), and in hot ethanol it decomposes reversibly—

$$2Ph_3GeSNa \rightleftharpoons (Ph_3Ge)_2S + Na_2S$$

Reactions with halogen compounds proceed in the expected manner—

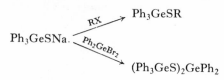

Compounds of the types R_3GeSR', $R_2Ge(SR')_2$ and $RGe(SR')_3$ are all readily obtained by straightforward reactions—

$$R_3GeX + R'SNa \longrightarrow R_3GeSR' + NaX$$

$$R_2GeBr_2 + 2R'SH \xrightarrow{C_5H_5N} R_2Ge(SR')_2 + C_5H_5NHBr$$

If these are carried out in benzene, then pyridinium bromide may be filtered directly from the reaction mixture (Davidsohn *et al.*, 1965; Mehrotra *et al.*, 1967b). Compounds of the type $R_2Ge(X)SR'$ have been obtained by equilibration reactions—

$$Me_2GeX_2 + Me_2Ge(SMe)_2 \underset{}{\overset{120°C}{\rightleftharpoons}} 2Me_2Ge(X)SMe$$

The course of the reaction is readily followed by the p.m.r. spectrum, and equilibrium constants have been evaluated where X = chlorine, bromine and iodine (Moedritzer and Van Wazer, 1965a). Similar exchange reactions have been used either to establish a Ge—SR bond, or to convert from one —SR group to another (Anderson, 1956b; Abel *et al.*, 1966)—

$$Me_3SiSR + Me_3GeBr \rightarrow Me_3GeSR + Me_3SiX$$
$$(R_3Ge)_2O + 2R'SH \rightarrow 2R_3GeSR' + H_2O$$
$$R_3GeSMe + R'SH \rightarrow R_3GeSR' + MeSH$$

Other methods that are not important from the preparative point of view include the reaction between hydrides and mercaptans, and the cleavage of Ge—NR_2 bonds by mercaptans—

$$Et_3GeH + BuSH \xrightarrow{Pt} Et_3GeSBu + H_2$$
$$R_3GeNR'_2 + RSH \longrightarrow R_3GeSR + R'_2NH$$

Some reactions of GeSR compounds have been reported. For example, aqueous silver nitrate gives the digermoxane, as do other oxidizing agents, such as H_2O_2. The germanium—sulphur bond is cleaved by n-butyl-lithium or Grignard reagents (Satge and Lesbre, 1965)—

$$2Ph_3GeSMe + 2AgNO_3 + H_2O \longrightarrow (Ph_3Ge)_2O + 2AgSMe + 2HNO_3$$

$$Ph_3GeSMe \begin{cases} \xrightarrow{Bu^nLi} Ph_3GeBu^n + MeSLi \\ \xrightarrow{PhMgX} Ph_4Ge + MeSMgX \end{cases}$$

A reaction of some interest is the cleavage of the germanium—sulphur bond by alkyl halides, which may proceed via a germylsulphonium salt $[R_3GeSMe_2]^+I^-$, but a careful study reveals only the final cleavage products (Hooton and Allred, 1965)—

$$Me_3GeSMe + 2MeI \rightarrow Me_3GeI + [Me_3S]^+I^-$$

Dimethyl sulphate cleaves the germanium—sulphur bond forming a sulphonium methyl sulphate—

$$Me_3GeSMe + (MeO)_2SO_2 \rightarrow Me_3GeOSO_2OMe + [Me_3S]^+[MeOSO_3]^-$$

Methyltribromogermane forms a more complex compound with hydrogen sulphide—

$$4MeGeBr_3 + 6H_2S + 12Et_3N \rightarrow Me_4Ge_4S_6 + 12Et_3NHBr$$

The structure has not been completely resolved, but two possibilities are (9) and (10) (Moedritzer, 1967), and analogous silicon compounds are known—

(9)   (10)

By contrast, germanoic anhydrides apparently yield sulphur analogues (Bauer and Burschkies, 1932)—

$$(RGeO)_2O + 3H_2S \xrightarrow{\text{HOAc}} (RGeS)_2S + 3H_2O$$

Compounds of the type $(R_3Ge)_2S$ and $(R_2GeS)_n$ have been prepared in aqueous or alcoholic media by the reaction—

$$2R_3GeX + Na_2S \rightarrow (R_3Ge)_2S + 2NaX$$
$$R_2GeX_2 + H_2S \rightarrow (R_2GeS)_n + 2HX$$

Sulphur is considerably more reactive towards germanium—carbon bonds than is oxygen, and one or two germanium—carbon bonds in organo-germanes are quite readily broken by reaction with sulphur at temperatures up to 250°C (Schmidt and Schumann, 1963). For example, tetra-n-butyl-germane gives the cyclic trimer, which is probably formed via the dibutyl-thiogermane—

$$Bu^n_4Ge \xrightarrow{S} Bu^n_2Ge(SBu^n)_2 \longrightarrow (Bu^n_2GeS)_3 + Bu^n_2S$$

Similarly one germanium—carbon bond in the germanacyclobutane ring is broken with insertion of sulphur (Mazerolles *et al.*, 1968)—

$$Bu^n_2Ge \diamondsuit \xrightarrow{\text{S, 250°C}} Bu^n_2Ge \overset{}{\underset{S}{}}$$

Arylgermanium bonds are even more susceptible to attack by sulphur, and tetraphenylgermane is completely degraded—

$$Ph_4Ge \xrightarrow{S} 2Ph_2S + Ge$$

Both tetraphenyl-tin and -lead behave differently—

$$Ph_4Pb \xrightarrow{S} PbS + 2Ph_2S$$

$$Ph_4Sn \xrightarrow{S} (Ph_2SnS)_3 + Ph_2S$$

The cyclic trimer $(Me_2GeS)_3$ and the silicon analogue equilibrate over about 2 weeks at 140°C, and at equilibrium four 6-membered ring systems are present. In addition, there exists an equilibrium between dimeric and trimeric silicon species—

$$3(Me_2SiS)_2 \rightleftharpoons 2(Me_2SiS)_3$$

P.m.r. data for this system have been evaluated in terms of three equilibrium constants (Moedritzer and Van Wazer, 1967).

Although various selenium– and tellurium–germanium compounds are known, few reactions or properties have been described (Vyazankin *et al.*, 1966d)—

$$Et_3GeH + Se \longrightarrow H_2Se + Et_3GeSeH + (Et_3Ge)_2Se$$

$$Et_3GeH + Et_2Se \xrightarrow{200°C} (Et_3Ge)_2Se + C_2H_6$$

Mixed compounds of the type $R_3GeSeSnR_3$ are decomposed by mineral acid, liberating hydrogen selenide, and by oxygen with the deposition of selenium. Cyclic compounds, such as $(Me_2GeSe)_3$, and the lithium derivatives, $Ph_3GeSeLi$ and $Ph_3GeTeLi$ have been prepared (Schmidt and Ruf, 1963; Ruidisch and Schmidt, 1963a; Schumann *et al.*, 1964b, 1965).

7 | ORGANOGERMANIUM HALIDES AND PSEUDOHALIDES

Germanium halides feature extensively in all aspects of organogermanium chemistry, and hence most of their chemical reactions are discussed in other chapters. Despite the range of known compounds (R_nGeX_{4-n}, $R_2Ge(H)X$, $RGe(H)_2X$, $R_nGe_2X_{6-n}$, $XGeR_2(GeR_2)_nR_2GeX$), there is a lack of information on the variation of reactivity with the organic groups present, or the particular halide. There has also been little comparative work on the reactivity of related silicon-, germanium- and tin-halides. The thermal stability of organogermanium halides is considerable; even ethyltri-iodo-germane is stable to 300°C, but above 350°C it decomposes to germanium tetraiodide and hydrocarbons (Flood, 1933).

I. PREPARATIVE METHODS

The many types of reactions that result in the formation of organo-germanium halides may be classified under the following headings.

A. Partial alkyl- or aryl-ation of germanium tetrahalides

The use of Grignard reagents in this connection is not usually a satis-factory reaction, since merely adjusting the stoicheiometry does not result in a high yield of a single product. There are however several examples where reasonable yields of mono-, di- or tri-halides have been obtained (Satge, 1961; Leusink *et al.*, 1964)—

$$\text{n-}C_6H_{13}MgCl + GeCl_4 \longrightarrow \text{n-}C_6H_{13}GeCl_3$$

$$2\text{n-}C_6H_{13}MgCl + GeCl_4 \longrightarrow (\text{n-}C_6H_{13})_2GeCl_2$$

$$PhMgBr + GeCl_4 \longrightarrow PhGeBr_3$$

$$Me_2GeCl_2 + BrMg\text{—}\langle\bigcirc\rangle\text{—}MgBr \longrightarrow Me_2(Cl)Ge\text{—}\langle\bigcirc\rangle\text{—}Ge(Cl)Me_2$$

Dimethyldimethoxygermane has been partially methylated, giving tri-methyliodogermane rather than the methoxide. Similar reactions in which

122

"mixed" halides are present frequently result in halide exchange (West *et al.*, 1954)—

$$Me_2Ge(OMe)_2 + MeMgI \rightarrow Me_3GeI$$

The Grignard method is more successful in producing R_3GeX compounds when organic groups having a large steric requirement are present, and there are several examples where even a large excess of the Grignard reagent fails to substitute the fourth halogen (West, 1952; Eaborn and Pande, 1960b)—

$$(1\text{-naphthyl})MgBr + (C_{10}H_8)_2GeCl_2 \rightarrow (C_{10}H_8)_3GeBr \quad 38\%$$

$$(cyclohexyl)MgBr + (C_6H_{11})_2GeBr_2 \rightarrow (C_6H_{11})_3GeBr \quad 50\%$$

t-Butylmagnesium chloride and germanium tetrachloride produce only low yields of Bu^tGeCl_3, and further alkylation is not observed. The comparatively low yields obtained even with bulky organic groups may be ascribed to a variety of side reactions, including the formation of germyl Grignard reagents, R_3GeMgX, and coupled products derived from them. Although most Grignard reactions are carried out in ethereal solvents, the reaction can be modified if hydrocarbons such as toluene or hexane are used. Thus n-butyl chloride and magnesium react exothermically with germanium tetrachloride, giving after 2 hrs under reflux, tri-n-butylchlorogermane in 60% yield (Zakharkin *et al.*, 1962).

Several examples have been reported in which an organolithium compound reacts more selectively than a Grignard reagent in forming organogermanium halides. For example, trimethylchlorogermane is obtained in quite high yield by the reaction (Schmidt and Ruidisch, 1961)—

$$Me_2GeCl_2 + MeLi \xrightarrow{0°C} Me_3GeCl$$

Steric considerations are probably responsible for the isolation of a carboranegermanium dihalide, which is formed via a lithio-derivative (Zakharkin *et al.*, 1965)—

Several less reactive organometallic compounds have been used for the partial alkyl- or aryl-ation of germanium tetrachloride. The reactions are of greater interest in reflecting relative reactivities than as preparative methods; thus mercury-alkyls and -aryls substitute only one chlorine atom (Orndorf *et al.*, 1927; Brinckman and Stone, 1959)—

$$R_2Hg + GeCl_4 \rightarrow RGeCl_3 + RHgCl \quad (R = Ph, CH_2:CH, etc.)$$

Tetraorgano-stannanes and -plumbanes have been used to monoalkylate germanium tetrachloride, and in individual cases high yields are obtained (Mironov and Kravchenko, 1964; Grant and Van Wazer, 1965; Neumann and Kuehlein, 1967b)—

$$Me_4Sn + HGeCl_3 \rightarrow MeGe(H)Cl_2 + Me_3SnCl$$

The rate of reaction of tetramethyltin with germanium tetrachloride is slow at 120°C in carbon tetrachloride and is complicated by reaction with the solvent, since methyl chloride is also formed. Tetraethyl-lead provides a most satisfactory route to ethyltrichlorogermane—

$$Et_4Pb + GeCl_4 \xrightarrow{100°C} Et_3PbCl + EtGeCl_3 \quad (90\%)$$

At higher temperature (140°C), lead alkyls substitute a second halogen, since triethyl-lead chloride then disproportionates, generating a further quantity of tetraethyl-lead—

$$2Et_3PbCl \xrightarrow{140°C} Et_2PbCl_2 + Et_4Pb$$

Functionally substituted tin alkyls are more reactive, and can be used to alkylate germanium tetrachloride successively (Adveeva *et al.*, 1966)—

$$R_3SnCH_2CO_2Me + GeCl_4 \rightarrow Cl_3GeCH_2CO_2Me + R_3SnCl$$

1. *Direct synthesis*

This method, as the name implies, involves the reaction between the metal and an organic halide. In silicon chemistry it is the standard method of obtaining alkylchlorosilanes on an industrial scale, and the main features of the reaction are paralleled by germanium.

Methyl chloride and germanium react at 340°C in the presence of copper to give a liquid mixture containing some 56% dimethyldichlorogermane, together with the trimethyl- and monomethyl-germanium chlorides. Some loss of hydrocarbon occurs (as methane, etc.), but in contrast to silicon there is evidently no chlorohydride (Me_2GeHCl) formed. The reaction will proceed in the absence of copper, but a considerably higher reaction temperature (460°C) is then needed (Rochow, 1947). The yield of methyltrichlorogermane is increased by operating at higher temperatures and with a greater proportion of copper present (Rochow *et al.*, 1951; Ponomarenko and Vzenkova, 1957). Germanium powder alone, dispersed over glass wool, at 510°–520°C can give up to 70% of methyltrichlorogermane (Wieber *et al.*, 1966). Similar copper-catalysed reactions using methyl bromide and ethyl chloride have been studied, and chlorobenzene requires a higher temperature and the presence of a silver catalyst when it gives predominantly diphenyldichlorogermane—

$$2PhCl + Ge/Ag \rightarrow Ph_2GeCl_2$$

Detailed conditions have been described for producing optimum yields of ethylchlorogermanes (Zueva *et al.*, 1966). Methylene chloride combines with germanium in the presence of copper giving a more complex mixture of products, three of which have been identified (Mironov and Gar, 1964a)—

$$CH_2Cl_2 + Ge/Cu \longrightarrow \underset{27\%}{CH_3GeCl_3} + \underset{23\%}{CH_2(GeCl_3)_2} + Cl_2Ge\begin{array}{c} Cl_2 \\ -Ge \\ \diagdown \\ -Ge \\ Cl_2 \end{array}$$

$$19\%$$

Similar reactions using alkenyl halides follow essentially the same course (Mironov *et al.*, 1961)—

$$CH_2:CHCH_2Br + Ge/Cu \xrightarrow{300°C} CH_2:CHCH_2GeBr_3$$

Somewhat related to the direct synthesis is the reaction between organic halides and germanium tetrahalides in the presence of copper powder whereby the organogermanium trihalide is produced. This simple preparative procedure has not yet been extensively explored (Mironov and Fedotov, 1966)—

$$1\text{-}C_{10}H_7I + GeBr_4 + Cu \xrightarrow{reflux} 1\text{-}C_{10}H_7GeBr_3 \quad (64\%)$$
$$PhI + GeCl_4 + Cu \longrightarrow PhGeI_3 \qquad\qquad (80\%)$$

2. *Cleavage of germanium—carbon bonds by halogens and hydrogen halides*

This method has mainly been employed to produce mono- and di-halides, but the reaction of chlorine with tetraorganogermanes has scarcely been investigated. Aryl groups are more readily cleaved by bromine or iodine than are alkyl groups.

Tetramethylgermane reacts quantitatively with bromine in isopropyl bromide (Mironov and Kravchenko, 1965) or rather less conveniently with neat bromine at room temperature. If a catalytic amount of aluminium bromide is added the reaction proceeds one stage further—

$$Me_4Ge + Br_2 \begin{cases} \xrightarrow{Pr^iBr} Me_3GeBr + MeBr & 98\% \\ \xrightarrow[AlBr_3]{Pr^iBr} Me_2GeBr_2 + 2MeBr & 80\% \end{cases}$$

Acetyl chloride in the presence of aluminium chloride will cleave either one or two methyl groups (Sakurai *et al.*, 1966)—

$$Me_4Ge + CH_3COCl \xrightarrow{AlCl_3} Me_3GeCl + Me_2GeCl_2$$

Tetraethylgermane was the first organogermane to be brominated and, in 1,2-dibromoethane or ethyl bromide, triethylbromogermane is produced in about 80% yield (Eaborn and Pande, 1960b)—

$$Et_4Ge + Br_2 \xrightarrow{C_2H_4Br_2} Et_3GeBr + EtBr$$

Early observations by Flood (1932) revealed a six-fold change in the relative rates of bromination of triethylgermanium halides in the order—

$$Et_3GeF > Et_3GeCl > Et_3GeBr$$

Side-chain bromination of triethylbromogermane must also occur, since appreciable amounts of hydrogen bromide are formed. Direct bromination of the higher tetra-alkyl germanes is sometimes more complex and can result in side-chain bromination without cleavage of germanium—carbon bonds. Tri-n-propylbromogermane is formed in high yield from equimolar amounts of bromine and tetra-n-propylgermane, but with excess bromine much hydrogen bromide is evolved owing to substitution in the alkyl groups, and compounds, such as $(C_3H_6Br)_3GeBr$, have been identified (Carrick and Glockling, 1966).

The direct chlorination of trimethylchlorogermane has been studied in solution and in the gas phase. Under both conditions, substitution reactions predominate, producing $Me_2(CH_2Cl)GeCl$ and $Me(CHCl_2)_2GeCl$. The gas-phase reaction can produce high yields (84%) of the monochloromethyl derivative (Wieber et al., 1967).

Use has been made of the more facile cleavage of phenyl groups as a way of obtaining dihalides (Flood, 1932)—

$$Ph_4Ge \xrightarrow{Br_2} Ph_2GeBr_2 \xrightarrow{EtMgBr} Ph_2GeEt_2 \xrightarrow{Br_2} Et_2GeBr_2$$

One n-propyl group is cleaved from tri-n-propylfluorogermane by bromine in the presence of iron powder (Anderson, 1952b)—

$$Pr^n_3GeF + Br_2 \rightarrow Pr^n_2GeBrF + Pr^nBr$$

The brominative cleavage of heterocyclic organogermanes shows some further features of interest. Four- or five-membered rings are cleaved in preference to alkyl groups, and 4-membered rings are the most reactive of all (Mazerolles, 1962; Mazerolles et al., 1966)—

Iodination of tetra-n-butylgermane in the absence of solvent or catalyst leads to the quantitative cleavage of one butyl group. If aluminium iodide is used as catalyst, then further reaction is possible; the fate of the alkyl group cleaved depends on its size. For tetramethyl- and tetraethyl-germanes it forms only the alkyl iodide, whereas tetra-n-propylgermane gives a mixture of n-hexane and n-propyl iodide. Tetra-n-butyl-, tetraisobutyl- and tetra-n-amyl-germanes give exclusively the hydrocarbon dimer (Lesbre and Mazerolles, 1958)—

$$R_4Ge + I_2 + AlI_3 \begin{cases} \longrightarrow R_3GeI + RI \\ \longrightarrow R_3GeI + R_2 \end{cases}$$

If triethyliodogermane is iodinated, the ethyl group forms both ethyl iodide and butane, although further cleavage of an ethyl group gives only butane—

$$Et_3GeI \xrightarrow{I_2, AlI_3} Et_2GeI_2 + EtI + C_4H_{10}$$

$$\downarrow I_2, AlI_3$$

$$EtGeI_3 + C_4H_{10}$$

Iodine chloride will also cleave the germanacyclobutane ring—

$$Bu^n_2Ge{\Large\diamondsuit} + ICl \longrightarrow Bu^n_2Ge{\Big\langle}^{Br}_{(CH_2)_3I}$$

The halogenation of organodigermanes can give almost exclusively the triorganohalogermane, and this is the most satisfactory method for the preparation of trivinylbromogermane (Seyferth, 1957)—

$$(CH_2:CH)_6Ge_2 + Br_2 \rightarrow (CH_2:CH)_3GeBr$$

Unsaturated organic groups bonded to germanium are frequently cleaved under extremely mild conditions, although in some cases brominative addition to the carbon—carbon double bond is observed (Satge, 1961; Sarankina and Manulkin, 1966)—

$$Bu^n_3GeCH_2CH:CH_2 \xrightarrow{Br_2, -80°C} Bu^n_3GeBr + CH_2:CHCH_2Br$$

$$Bu^n_3GeCH:CH_2 \xrightarrow{Br_2} Bu^n_3GeCHBrCH_2Br$$

$$Bu^n_3GeCH:CHPh \xrightarrow{Br_2} Bu^n_3GeBr + PhCH:CHBr$$

$$Ph_2Ge(CH:CHMe)_2 \xrightarrow[40°C]{Br_2, CHCl_3} Ph_2GeBr_2$$

Where cleavage rather than addition occurs under mild conditions, it is tempting to postulate a primary addition reaction followed by β-elimination—

$$R_3GeCH_2CH:CH_2 \longrightarrow R_3Ge \overset{CH_2}{\underset{Br}{\diagdown}} CHCH_2Br \longrightarrow$$

$$R_3GeBr + CH_2:CHCH_2Br$$

Other organogermanes with substituents in the β position are readily cleaved by bromine—

$$Bu^n{}_3GeCH_2CO_2Me \xrightarrow{Br_2} Bu^n{}_3GeBr + BrCH_2CO_2Me$$

Aluminium halides, in the presence of alkyl halides, catalyse the cleavage of one organic group from a tetraorganogermane. This forms an extremely useful preparative method that clearly has features in common with the redistribution reaction. When tetraethylgermane is boiled with isopropyl-chloride containing 2% of aluminium chloride, one ethyl group is removed and a mixture of hydrocarbons (C_2H_4, C_2H_6, C_3H_8) has been isolated in addition to quite high yields of triethylchlorogermane. Chlorides, bromides and iodides have been obtained in this way as in the following examples (Mironov and Kravchenko, 1965)—

$$Me_4Ge + Pr^iCl \xrightarrow{AlCl_3} Me_3GeCl \quad 95\%$$

$$Bu^n{}_4Ge + i\text{-}C_5H_{11}Br \xrightarrow{AlBr_3} Bu^n{}_3GeBr \quad 67\%$$

$$Me_3GePh + Bu^nI \xrightarrow{AlCl_3} Me_3GeI \quad 80\%$$

Brominative cleavage of either one or two phenyl groups from tetra-phenylgermane occurs readily, and provides a straightforward preparative route to both Ph_3GeBr and Ph_2GeBr_2 (Johnson et al., 1957). Phenyl-germanes are more readily cleaved than either alkyl groups or cyclic systems, as illustrated by the following example (Mazerolles et al., 1966)—

Tetrabenzylgermane is readily brominated to tribenzylbromogermane, but with an excess of bromine, hydrogen bromide is evolved as polybrominated products are formed (Cross and Glockling, 1964). One phenyl group has been cleaved from tetraphenylgermane by iodine in refluxing decahydronaphthalene (Manulkin *et al.*, 1963), whereas the carbinol, $Ph_3GeC(OH)Ph_2$ undergoes what is probably a carbonium ion rearrangement when treated with boron trifluoride etherate (Brook *et al.*, 1964)—

$$Ph_3GeC(OH)Ph_2 \xrightarrow{F_3BOEt_2} Ph_2Ge(F)CPh_3$$

Trialkylfluorogermanes are conveniently prepared by the action of hydrogen fluoride on a tetra-alkylgermane (Gladshtein *et al.*, 1959)—

$$R_4Ge + HF \rightarrow R_3GeF + RH$$

Hydrogen chloride and hydrogen bromide cleave tetramethylgermane and, although yields are low, there is no evidence that more than one methyl group is removed (Dennis and Patnode, 1930; Griffiths and Onyszchuk, 1961). The reaction is usually more satisfactory if aluminium-chloride or -bromide is added. Benzyl and aryl groups are more readily cleaved by halogen acids, and comparative rate measurements using hydrogen bromide in chloroform at room temperature have demonstrated that p-tolyl groups are most readily cleaved; reaction rates decrease in the order p-tolyl > m-tolyl > phenyl > benzyl (Simons, 1935). The differences in rate are sufficiently great for unsymmetrical germanes to be selectively cleaved—

$$Ph_3Ge(C_6H_4Me\text{-}m) + HBr \rightarrow Ph_3GeBr + PhMe$$

Hydrogen bromide can cleave two groups from tetra-m-tolylgermane whereas, under similar conditions, only one group is cleaved from tetra-o-tolylgermane. Kinetic evidence suggests that the acid cleavage of aryl—germanium bonds is best regarded as an electrophilic substitution at carbon rather than nucleophilic substitution at germanium. Germanacyclanes are readily cleaved by hydrogen halides (Mazerolles *et al.*, 1966)—

$$Bu^n_2Ge\text{—cyclobutane} + HX \longrightarrow Bu^n_2Ge(X)Pr^n$$

The reactions of unsaturated organogermanes with hydrogen bromide appear to depend on the particular set of experimental conditions. Thus, allyl groups are readily cleaved in what may be an addition reaction followed by β-elimination of R_3GeBr—

$$Bu^n_3GeCH_2CH{:}CH_2 + HBr \rightarrow Bu^n_3GeCH_2CH(Br)CH_3$$

$$\downarrow$$

$$Bu^n_3GeBr + CH_3CH{:}CH_2$$

5*

However, vinyl groups appear to add hydrogen bromide—

$$Bu^n_3GeCH:CH_2 + HBr \rightarrow Bu^n_3GeCH_2CH_2Br$$

β-Elimination reactions from compounds of this type often require the presence of a base, and are in competition with the elimination of hydrogen halide (Mironov and Dzhurinskaya, 1963)—

$$ClCH_2CH_2GeCl_3 \xrightarrow{\text{piperidine}} CH_2:CHGeCl_3 + GeCl_4$$

$$ClCH_2CHClGeCl_3 \xrightarrow{\text{quinoline}} CH_2:CHClGeCl_3 + GeCl_4$$

The isomeric chlorovinyltrichlorogermane is obtained by a dehydrochlorination reaction catalysed by aluminium trichloride—

$$ClCH_2CHClGeCl_3 \xrightarrow{\text{AlCl}_3} CHCl:CHGeCl_3 + HCl$$

Trichlorogermyl-allyl and -vinyl compounds seem less prone to either addition or cleavage by hydrogen bromide and may react simply by halide exchange (Mironov et al., 1961)—

$$Cl_3GeCH:CH_2 + 3HBr \rightarrow Br_3GeCH:CH_2 + 3HCl$$

B. Organogermanium oxides and hydrogen halides

Several of the more important preparative methods discussed earlier produce a mixture of halides, R_3GeX, R_2GeX_2 and $RGeX_3$, and separation by distillation or fractional crystallization presents difficulties, especially for the lower alkylchlorogermanes, which tend to form azeotropes. In these cases, it has been common practice to hydrolyse a mixture with aqueous ammonia or sodium hydroxide, separate the oxides and then reconvert the pure oxides to the corresponding halides with strong aqueous mineral acid. Many examples have been reported; thus, bistri-n-propylgermanium oxide reacts exothermically with all the halogen acids to give quantitative yields of the halides. Germoxanes containing germanium—hydrogen bonds are converted into halohydrides (Kraus and Brown, 1930a; Anderson, 1951a; Fuchs and Gilman, 1958; Massol and Satge, 1966)—

$$(Pr^n_3Ge)_2O + 2HX \rightarrow 2Pr^n_3GeX + H_2O$$

$$[EtGe(H)_2]_2O + 2HF \rightarrow 2EtGe(H)_2F + H_2O$$

In the same way, cyclic or polymeric oxides of the type $(R_2GeO)_n$ are converted, usually quantitatively, into the dihalides (Anderson, 1953), whereas polymeric germanoic anhydrides have been used to prepare trihalides—

$$(R_2GeO)_n \xrightarrow{HX} R_2GeX_2 + H_2O$$

$$[(RGeO)_2O]_n \xrightarrow{HX} RGeX_3 + H_2O$$

1. *Halogenation of organogermanium hydrides*

Reactions of this type are of greatest use in forming organogermanium halohydrides, and are of interest in connection with the mechanism of the substitution. Direct halogenation proceeds under mild conditions and, with iodine, may even be controlled so as to replace only one hydrogen atom in a di- or tri-hydride (Anderson, 1960)—

$$Bu^n_2GeH_2 \xrightarrow{I_2} Bu^n_2Ge(H)I + H \quad I(57\%)$$

$$Bu^nGeH_3 \xrightarrow{I_2} Bu^nGe(H)_2I + HI \quad (70\%)$$

When applied to triorganogermanes, the reaction is usually quantitative at room temperature for chlorine, bromine and iodine. However, experimental conditions can profoundly alter the course of the reaction. It is, for example reported that, whereas bromine and diphenylgermane combine smoothly to give diphenyldibromogermane, the corresponding reaction with iodine produces at 70°C a vigorous secondary reaction resulting in cleavage of the phenyl—germanium bonds, so that the final product is germanium tetra-iodide (Johnson and Harris, 1950). Several halogenation reactions have been reported on asymmetric organogermanium hydrides, though without experimental detail (p. 65). Whereas *N*-bromosuccinimide at 0°C results in racemization, chlorine does not—

$$(+)MePh(1\text{-}C_{10}H_7)GeH \xrightarrow{Cl_2} (+)MePh(1\text{-}C_{10}H_7)GeCl$$

$$\downarrow LiAlH_4$$

$$(-)MePh(1\text{-}C_{10}H_7)GeH$$

With an excess of halogen, both di- and tri-hydrides are fully halogenated (Anderson, 1960), whereas the digermoxane, $(Et_2GeH)_2O$, is converted into diethyldibromogermane by bromine—

$$R_2GeH_2 \xrightarrow{I_2} R_2GeI_2 \qquad RGeH_3 \xrightarrow{Br_2} RGeBr_3$$

$$(Et_2GeH)_2O \xrightarrow{Br_2} 2Et_2GeBr_2 + H_2O$$

Mercuric halides have been used to monohalogenate organogermanes; these reactions take place at room temperature, and are essentially quantitative. The chlorination of ethylgermane may be taken one stage further using chloromethyl ether. *N*-Bromosuccinimide and *N*-iodosuccinimide also monohalogenate ethylgermane (Massol and Satge, 1966)—

$$Bu^n_2GeH_2 \xrightarrow{HgX_2} Bu^n_2Ge(H)X$$

$$EtGeH_3 \xrightarrow{HgCl_2} EtGe(H)_2Cl \xrightarrow[AlCl_3]{ClCH_2OMe} EtGe(H)Cl_2 + Me_2O$$

In some cases, chloromethyl ether combines without having aluminium chloride present, and alkyl halides undergo the same reaction at reflux temperature (Satge, 1961)—

$$Et_2GeH_2 \xrightarrow{EtI} Et_2Ge(H)I + C_2H_6$$

$$Bu^n_2GeH_2 \xrightarrow{RCl} Bu^n_2GeCl_2 + 2RH$$

Many other halogen-containing compounds halogenate organogermanes, often in exothermic reactions (Satge, 1961). Germanium tetrachloride, carbon tetrachloride, sulphuryl chloride, acid chlorides, trichloroacetic acid and chloroacetone are all effective halogenating agents—

$$R_3GeH \begin{cases} \xrightarrow{GeCl_4,\ Et_2O} R_3GeCl + HGeCl_3 . 2Et_2O \\ \xrightarrow{SO_2Cl_2} R_3GeCl + SO_2 + HCl \\ \xrightarrow{PhCOCl} R_3GeCl + PhCHO \end{cases}$$

A similar "redistribution" reaction occurs between trialkylgermanes and dialkyldihalogermanes, and these are exothermic in the presence of aluminium halide—

$$R_3GeH + R_2GeX_2 \xrightarrow{AlX_3} R_2Ge(H)X + R_3GeX$$

Highly selective free-radical bromination reactions have been carried out on phenyl- and diphenyl-germanes (Satge and Riviere, 1966)—

$$PhGeH_3 \xrightarrow{N\text{-bromosuccinimide}} PhGe(H)_2Br$$

$$Ph_2GeH_2 \xrightarrow{N\text{-bromosuccinimide}} Ph_2Ge(H)Br$$

Hydrogen chloride in the presence of aluminium trichloride reacts with di- and tri-hydrides to give mixtures of products (Amberger and Boeters, 1961)—

$$MeGeH_3 \xrightarrow[AlCl_3]{HCl} MeGe(H)_2Cl + MeGe(H)Cl_2 + MeGeCl_3 + H_2$$

2. Redistribution reactions

The term "redistribution" may be applied to any reaction in which organic groups originating on one metal are transferred to another. It is customary, in practice, to restrict the term to reactions in which only one metal is involved, and the most common type is that between a metal halide and an organo-derivative of the same metal—

$$R_4M + MCl_4 \rightleftharpoons R_3MCl + R_2MCl_2 + RMCl_3$$

These equilibria reactions are common to silicon, germanium, tin and lead (Moedritzer, 1966a). The majority have been studied under non-equilibrium conditions, but in the past few years there have been several investigations, usually by p.m.r. techniques, carried out under equilibrium conditions. The greater the metal—carbon bond strength the more drastic are the conditions needed; tin compounds undergo redistribution reactions without added catalyst under quite mild conditions. Arylgermanes will also redistribute without added catalyst, but only at high temperature, whereas alkylgermanes for the most part require a catalyst (such as an aluminium halide or germanium di-iodide).

Tetraphenylgermane and germanium tetrachloride when heated in a sealed tube react mainly according to the equation—

$$Ph_4Ge + GeCl_4 \xrightarrow{250°C} PhGeCl_3 \quad (75\%)$$

with aluminium chloride present, the reaction temperature is reduced to 120°C (Schwarz and Schmeisser, 1936; Rijkens and Van der Kerk, 1964). Many reactions of this type have been examined, and a fairly exhaustive study has been made of the reaction between tetra-n-butylgermane and germanium tetrachloride (Mazerolles, 1961; Van der Kerk $et\ al.$, 1962; Luijten and Rijkens, 1964; Rijkens $et\ al.$, 1966). The yields of products are governed mainly by the stoicheiometry and reaction temperature, but a mixture is always produced—

$$Bu^n_4Ge + GeCl_4 \xrightarrow[200°C]{GeI_2} Bu^n_3GeCl + BuGeCl_3 + trace\ GeCl_4$$

In the absence of a catalyst, tetra-n-butylgermane does not react at 200°C, but hexabutyldigermane and germanium tetrachloride react cleanly to give the pentabutyl compound—

$$Bu^n_6Ge_2 + GeCl_4 \rightarrow Bu^n_3GeGe(Cl)Bu^n_2$$

Mixtures of organogermanium halides are difficult to separate by traditional methods, and it has been common practice to effect purification by hydrolysis, separation of the mixture of organogermanium oxides followed by reconversion to the halides. With the general introduction of preparative-scale vapour-phase chromatography, direct separation of R_3GeX, R_2GeX_2 and $RGeX_3$ mixtures is much simpler, and this makes redistribution reactions more attractive as preparative methods.

The role of the catalyst in these reactions is not clearly understood; most studies have been made with aluminium halides, which probably de-alkylate the tetra-alkylgermane, forming alkylaluminium halides, which are the reactive intermediates in the equilibration processes. The de-alkylating -alkylating action of aluminium bromide has been demonstrated for hexaethyldigermane where, at 200°C, tetraethylgermane is formed, hence both the germanium—germanium and a germanium—carbon bond must be cleaved (Vyazankin *et al.*, 1964b)—

$$Et_6Ge_2 + AlBr_3 \rightarrow Et_4Ge$$

A redistribution reaction has been reported on a germanacyclobutane, and sulphuryl chloride chlorinates the heterocyclic ring (Mazerolles *et al.*, 1966)—

$$Bu^n_2Ge \diamondsuit \quad \begin{array}{l} \xrightarrow{GeCl_4} \quad Bu^n_2Ge \diagdown^{Cl}_{(CH_2)_3GeCl_3} \\ \xrightarrow{SO_2Cl_2} \quad Bu^n_2Ge \diagdown^{Cl}_{(CH_2)_3Cl} \end{array}$$

Gallium trichloride is a most effective catalyst for redistribution reactions (Schumann *et al.*, 1967)—

$$Me_4Ge + Me_2GeCl_2 \xrightarrow[165°C]{GaCl_3} 2Me_3GeCl \quad (100\%)$$

$$3Et_4Ge + GeCl_4 \xrightarrow[150°C]{GaCl_3} 4Et_3GeCl$$

3. *Synthesis from halogermanium hydrides*

The versatility of trihalogermanes in the formation of organogermanium halides is quite striking. Most reactions have been studied with trichlorogermane, which reacts with all types of carbon—carbon double and triple bonds. The reaction with alkynes can readily be controlled, so that either one or two molecules of trichlorogermane are added. There are also many examples in which chlorogermanes have been added to conjugated systems of both double and triple bonds. Conjugated dienes react primarily by 1,4-addition and alkynes are more reactive than alkenes. Many of these reactions proceed under mild conditions without added catalyst, although free-radical (e.g., benzoyl peroxide) and ionic catalysts such as chloroplatinic acid have been used.

Trichlorogermane also reacts with various other compounds. Unsaturated organic halides combine either by addition or elimination of hydrogen halide, depending on the experimental conditions (Mironov *et al.*, 1962a; Lesbre *et al.*, 1963)—

$$CH_2 : CHCH_2Cl + HGeCl_3 \quad \begin{array}{l} \xrightarrow{ether} \quad CH_2 : CHCH_2GeCl_3 + HCl \\ \phantom{\xrightarrow{ether}} \quad 26\% \\ \xrightarrow{neat} \quad Cl(CH_2)_3GeCl_3 \\ \phantom{\xrightarrow{neat}} \quad 52\% \end{array}$$

Methallyl chloride combines in both ways, but related bromides and iodides appear to react only by elimination of hydrogen halide. Alkyl halides combine with trichlorogermane and this may provide the most general synthetic method for organotrichlorogermanes: even p-dichlorobenzene will react under forcing conditions (Mironov and Gar, 1965b)—

$$HGeCl_3 \begin{cases} \xrightarrow{Bu^tCl} Bu^tGeCl_3 & (40\%) \\ \xrightarrow{p\text{-}ClC_6H_4Cl} Cl-\langle\!\!\!\bigcirc\!\!\!\rangle-GeCl_3 & (11\%) \\ \xrightarrow{PhCH_2Cl} PhCH_2GeCl_3 & (45\%) \end{cases}$$

Trichlorogermane is sufficiently reactive to cleave ethers and substituted cyclopropanes (Nefedov et al., 1965a)—

$$R_2O + HGeCl_3 \longrightarrow RGeCl_3 + ROH$$

$$Ph-\triangle + HGeCl_3 \longrightarrow EtCH(Ph)GeCl_3 \qquad (85\%)$$

Tetra-alkylstannanes and plumbanes evolve some alkane with trichlorogermane, yielding a mixture of products (Mironov and Kravchenko, 1964)—

$$Me_4Sn + HGeCl_3 \rightarrow CH_4 + MeGe(H)Cl_2 + MeGeCl_3$$

Mono-olefins (C_5–C_{10}) have yielded alkyltrichlorogermanes in 20–30% yield (Fischer et al., 1954; Mironov and Gar, 1965a, 1966)—

$$RCH:CH_2 + HGeCl_3 \xrightarrow[50°-100°C]{(PhCO)_2O_2} RCH_2CH_2GeCl_3$$

Cyclic olefins react even more readily—

$$\bigcirc\!\!\!| \; + HGeCl_3 \longrightarrow \bigcirc\!\!\!-GeCl_3$$

$$\bigcirc\!\!\!| \; + HGeCl_3 \longrightarrow \bigcirc\!\!\!-GeCl_3$$

The addition of trichloro- or tribromo-germane to ethylene is more complex and the products isolated depend on the reaction temperature, but it seems likely that hydrogen and hydrogen chloride are also formed—

$$C_2H_4 + HGeCl_3 \longrightarrow Cl_3GeCH_2CH_2GeCl_3 + Cl_3GeCH_2CH_2Ge(H)Cl_2$$

$$C_2H_4 + HGeBr_3 \xrightarrow{50°C} Br_3GeCH_2CH_2GeBr_3 \qquad EtGeBr_3 + GeBr_4$$

$$\underset{> 50°C}{\underline{\hspace{8cm}}} \uparrow$$

These complexities are absent from the reaction between ethylene and methyldichlorogermane, and many reactions of this type proceed smoothly at temperatures between 100° and 150°C without added catalyst (Nefedov *et al.*, 1965b)—

$$MeGe(H)Cl_2 + C_2H_4 \rightarrow Me(Et)GeCl_2 \qquad (90\%)$$
$$Bu^n_2Ge(H)Cl + Me(CH_2)_7CH:CH_2 \rightarrow Bu^n_2(C_{10}H_{21})GeCl \ (80\%)$$

Butadiene and other conjugated dienes almost always give a mixture of products with trichloro- or tribromo-germane. Both 1,4-addition to butadiene occurs as well as the formation of a germanacyclopentene resulting from the addition of germanium dichloride to the diene (Mironov and Gar, 1966)—

$$MeCH:CHCH_2GeCl_3 \xleftarrow{C_4H_6} HGeCl_3 \rightleftharpoons$$

$$HCl + GeCl_2 \xrightarrow{C_4H_6} \underset{\substack{Ge \\ Cl_2}}{\square}$$

Dialkylchlorogermanes react more simply, giving only the 1,4-addition product, and there is chromatographic evidence that *cis–trans* isomers are produced (Satge and Massol, 1965)—

$$Et_2Ge(H)Cl + C_4H_6 \rightarrow Et_2Ge(Cl)CH_2CH:CHMe$$

Whereas benzene, toluene and the xylenes do not react with trichlorogermane, naphthalenes, anthracenes and phenanthracene form addition products. Trichlorosilane by contrast is unreactive. The only example that has been thoroughly studied is naphthalene, where a rapid reaction occurs at 110°C (Kolesnikov and Nefedov, 1965)—

Unlike simple ethers, anisole combines additively with trichlorogermane, and methylation of the complex crude reaction mixture has resulted in the isolation of a tris(trimethylgermyl) adduct—

$$\text{PhOMe} \xrightarrow[\text{2. MeMgBr}]{\text{1. HGeCl}_3}$$

Halogen-substituted olefins add trichlorogermane under mild conditions (Mironov and Gar, 1965b)—

$$\text{ClCH:CH}_2 + \text{HGeCl}_3 \rightarrow \text{ClCH}_2\text{CH}_2\text{GeCl}_3$$
$$\text{ClCH:CHMe} + \text{HGeCl}_3 \rightarrow \text{ClCH}_2\text{CH(CH}_3)\text{GeCl}_3$$

In many such reactions, the direction of addition to unsymmetrical double bonds has been deduced from the products of alkaline hydrolysis and particularly from the ionic halide produced. For example, in the presence of a base, the compound $\text{ClCH}_2\text{CH}_2\text{GeCl}_3$ undergoes a β-elimination reaction, and hence produces four titratable chloride ions—

$$\text{ClCH}_2\text{CH}_2\text{GeCl}_3 \xrightarrow{\text{NaOH}} \text{C}_2\text{H}_4 + \text{GeCl}_4 \xrightarrow{\text{NaOH}} \text{Na}_2\text{GeO}_4 + 4\text{NaCl}$$

If the addition of trichlorogermane to vinyl chloride had proceeded in the opposite direction giving $\text{CH}_3\text{CH(Cl)GeCl}_3$, then alkaline hydrolysis would have given only three moles of sodium chloride.

Organohalogermanium hydrides add to the carbon—carbon double bonds of a variety of substituted mono-olefins, and their reactivity is considerably greater than that of the simple organogermanium hydrides. Diethylchlorogermane combines with vinyl acetate even at room temperature (Satge *et al.*, 1966)—

$$\text{Et}_2\text{Ge(H)Cl} + \text{CH}_2\text{:CHOCOMe} \rightarrow \text{Et}_2\text{Ge(Cl)(CH}_2)_2\text{OCOMe}$$

Allyl alcohol is less reactive, but addition occurs without added catalyst (Satge, 1961)—

$$\text{Bu}^n_2\text{Ge(H)Cl} + \text{CH}_2\text{:CHCH}_2\text{OH} \xrightarrow{150°\text{C}} \text{Bu}^n_2[\text{HO(CH}_2)_3]\text{GeCl}$$

Acetone and other ketones react with trichlorogermane, but in a more complicated way, giving first the β-unsaturated ketone (Gar and Mironov, 1966; Nefedov *et al.*, 1967)—

$$\text{Me}_2\text{CO} \xrightarrow{\text{HGeCl}_3} \text{Me}_2\text{C:CHCOMe} \xrightarrow{\text{HGeCl}_3} \text{Cl}_3\text{GeC(Me)}_2\text{CH}_2\text{OMe}$$

Benzophenone behaves differently, since it first condenses to an ether, which is then cleaved by trichlorogermane

$$\text{Ph}_2\text{CO} \longrightarrow \text{Ph}_2\text{CHOH} \longrightarrow (\text{Ph}_2\text{CH})_2\text{O} \xrightarrow{\text{HGeCl}_3}$$
$$\text{Ph}_2\text{CHGeCl}_3 + \text{Ph}_2\text{CHOH}$$

An excess of trichlorogermane combines with acetylene to give 1,2-bis-trichlorogermylethane, but by adding trichlorogermane to a solution of acetylene in hexane, it is possible to obtain vinyltrichlorogermane in about 15% yield—

$$HGeCl_3 + C_2H_2 \rightarrow CH_2:CHGeCl_3 + Cl_3GeCH_2CH_2GeCl_3$$

In ether solution, both trichloro- and tribromo-germane react with acetylene to give olefinic addition products (Mironov and Gar, 1964b, 1965a)—

$$(Et_2O)_2HGeCl_3 + C_2H_2 \rightarrow CH_2:CHGeCl_3 + Cl_3GeCH:CHGeCl_3 + polymer$$

Trimethylethynylgermane produces both mono- and di-addition products with trichlorogermane, and mono-addition to vinylacetylene has been observed (Mironov *et al.*, 1964a; Mironov and Gar, 1965a, b)—

$$Me_3GeC:CH + HGeCl_3 \rightarrow Me_3GeCH:CHGeCl_3 + Me_3GeCH_2CH(GeCl_3)_2$$
$$CH_2:CHC:CH + HGeCl_3 \rightarrow CH_2:CHCH:CHGeCl_3$$

Dialkylchlorogermanium hydrides add to acetylene or substituted acetylenes at elevated temperatures (Lesbre *et al.*, 1963)—

$$Et_2Ge(H)Cl + C_2H_2 \xrightarrow{110°C} Et_2Ge(Cl)CH_2CH_2(Cl)GeEt_2$$
$$Bu^n{}_2Ge(H)Cl + PhC:CH \longrightarrow Bu^n{}_2Ge(Cl)CH:CHPh \qquad (75\%)$$

4. *From germanium dihalides*

Germanium di-iodide and the lower alkyl iodides react almost quantitatively according to the equation (Lesbre *et al.*, 1963)—

$$GeI_2 + RI \rightarrow RGeI_3$$

Methyl- and ethyl-iodide combine at 110°C in a sealed tube, but the temperature is somewhat critical since, with ethyl iodide at 140°C the main product is germanium tetraiodide (Flood, 1933). Trifluoromethyl iodide is exceptional, in that some $(CF_3)_2GeI_2$ is formed in addition to trifluoromethyltri-iodogermane (Clark and Willis, 1962).

A modification of this reaction has been described using a complex iodide such as $CsGeI_3$ (Tchakirian, 1939), and the corresponding chloride provides a versatile route to organotrichlorogermanes. Even t-butyltrichlorogermane can be obtained in this way as a white sublimable solid (Poskozim, 1968)—

$$CsGeCl_3 + Bu^tCl \rightarrow Bu^tGeCl_3 + CsCl$$

Mercury alkyls evidently combine with germanium di-iodide with deposition of mercury, although the only product reported is the digermane, $(Bu^n{}_2GeI)_2$, which has not been fully characterized (Jacobs, 1954). Mercury diaryls have been more thoroughly studied and the main reaction may be represented as—

$$Ar_2Hg + GeI_2 \rightarrow Ar_2GeI_2 + Hg \quad (40–75\%)$$

However other compounds are formed, including mercurous iodide, mercuric iodide and arylmercuric iodide, together with GeI_4, $ArGeI_3$ and Ar_3GeI. The formation of these products has been interpreted in terms of four competing reactions (Emel'yanova $et\ al.$, 1962)—

$$Ar_2Hg + Ar_2GeI_2 \rightarrow Ar_3GeI + ArHgI$$
$$ArHgI + GeI_2 \rightarrow ArGeI_3 + Hg$$
$$ArHgI + ArGeI_3 \rightarrow Ar_2GeI_2 + HgI_2$$
$$2HgI_2 + GeI_2 \rightarrow GeI_4 + Hg_2I_2$$

The reactions between germanium dihalides and alkynes or alkenes whereby germanium-containing heterocyclic compounds are produced have been discussed earlier (Chapter 3). Acetylene and germanium di-iodide yield, in addition to the 6-membered heterocyclic ring, benzene-soluble polymeric material having the composition $(C_2H_2GeI_2)_n$. Methylation of this material gives a polymer with an average molecular weight of about 33,000, for which a linear structure has been proposed (Vol'pin $et\ al.$, 1963)—

$$\left[CH:CHGe(Me)_2CH:CHGeMe_2 \right]_n$$

C. Germanium halides and diazo-compounds

Germanium tetrahalides in common with many other metal halides react readily and quantitatively with diazomethane and some of its substituted derivatives in the presence of copper powder (Kramer and Wright, 1963)—

$$GeX_4 + CH_2N_2 \xrightarrow[-60°C]{Cu} X_3GeCH_2X + N_2$$

The reaction proceeds more rapidly with germanium- than with silicon-halides, and if an excess of diazomethane is used, some disubstitution occurs. Phenyltrichlorogermane likewise gives a mixture of mono- and di-substitution products—

$$PhGeCl_3 + CH_2N_2 \xrightarrow{Cu} PhGe(Cl)_2CH_2Cl + PhGe(Cl)(CH_2Cl)_2 + N_2$$

In this case monosubstitution will occur without added copper and an excess of diazomethane in the presence of copper gives the disubstitution product quantitatively. Diphenyldichlorogermane undergoes a similar reaction, and aryldiazonium fluoroborates will monoarylate germanium tetrachloride, though in poor yield (Nesmeyanov $et\ al.$, 1958)—

$$Ph_2GeCl_2 + CH_2N_2 \longrightarrow Ph_2Ge(Cl)CH_2Cl + N_2$$
$$GeCl_4 + PhN_2BF_4 \xrightarrow{Zn} PhGeCl_3$$

Two mechanisms have been proposed for these reactions. Insertion of free methylene into the metal—X bond seems less likely than nucleophilic attack by diazomethane at the metal, followed by displacement of halide either as a stepwise or concerted process—

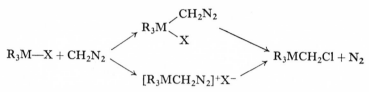

This type of mechanism finds support in a kinetic study on the reaction between *para*-substituted phenyltrichlorogermanes and diazomethane which shows that electron-withdrawing substituents (Cl, F) increase the rate, whereas electron-donating substituents (MeO) retard the reaction. Powdered copper has little if any effect on reaction rates (Seyferth and Hetflejs, 1968).

II. REACTIONS

A. Halide exchange

Several examples of halide exchange have already been encountered. These are all equilibria reactions, and by using an excess of any halide the required organogermanium halide can be isolated, usually in high yield. Fluorides have been obtained using antimony trifluoride or silver fluoride, and sodium iodide in acetone converts other halides into the iodides (Ponomarenko and Vzenkova, 1957; Clark and Willis, 1962)—

$$Me_3GeCl \xrightarrow{SbF_3} Me_3GeF$$

$$F_3CGeI_3 \xrightarrow{AgCl} F_3CGeCl_3$$

Tricyclohexylbromogermane is the main product of the reaction between cyclohexylmagnesium bromide and germanium tetrachloride (Bauer and Burschkies, 1932). In reactions of this type, it is advisable to avoid the use of different halides.

B. Hydrolysis

Most aspects of the hydrolysis of organogermanium halides are discussed in Chapter 6; they are much less readily hydrolysed than organochlorosilanes, but more rapidly than tin analogues. The kinetics of nucleophilic substitution reactions at silicon show that these processes have the characteristics of S_N2 reactions, although they may equally well be interpreted in terms of an unstable pentaco-ordinate intermediate.

Johnson and Schmall (1958) studied the kinetics of hydrolysis of several triarylgermanium halides, but their interpretation has been severely criticized by Chipperfield and Prince (1963). The latter authors measured the rates of hydrolysis of triphenyl-, tri-p-tolyl- and tri-isopropyl-chlorogermane in aqueous acetone by following the conductivity change under conditions suitable for fast reactions—

$$R_3GeCl + H_2O \rightleftharpoons R_3GeOH + H^+ + Cl^-$$

Equilibrium constants at 25°C depend on the R groups (R = Ph, $2 \cdot 2 \times 10^{-6}$; p-tolyl, $4 \cdot 5 \times 10^{-6}$; Pr^i, $1 \cdot 15 \times 10^{-7}$). Under these conditions, chlorosilanes are completely hydrolysed. For each compound, the rate is first order with respect to the germyl halide over the concentration range, 10^{-4} to 10^{-2} M, the half-life varying between $0 \cdot 015$ and 20 sec. The conclusion reached in this study is that the mechanism probably involves rearward attack by water, either as an S_N2 reaction or through a pentaco-ordinate intermediate—

$$(H_2O)_n + R_3GeCl \longrightarrow \left[\begin{array}{c} R \quad R \\ \diagdown \diagup \\ H{-}O{-}{-}{-}Ge{-}{-}{-}Cl \\ \vdots \quad | \\ (H_2O)_{n-1}H \quad R \end{array} \right] \longrightarrow$$

$$R_3GeOH + [(H_2O)_{n-1}H]^+ + Cl^-$$

It has also been observed (Kraus and Flood, 1932) that triethylfluorogermane is less rapidly hydrolysed than the other triethylgermyl halides.

There is no evidence that any organogermanium halides hydrolyse by an S_N1 mechanism. In fact, sterically congested halides, such as tri-isopropylchlorogermane, are less readily hydrolysed than others; this is also compatible with an S_N2 type of reaction. Two isobutyliodogermanes formed in the reaction—

$$Bu^i_3Al + GeI_2 \rightarrow Bu^i_3GeGe(I)Bu^i_2 + Bu^i_7Ge_3I$$

are sufficiently unreactive to survive hydrolysis of the reaction mixture (Glockling and Light, unpublished work).

Two products are produced by the addition of trimethylchlorogermane to the substituted phosphine methylene, $Me_3SiCH:PMe_3$ (Schmidbaur and Tronich, 1967)—

$$Me_3SiCH:PMe_3 + Me_3GeCl \rightarrow Me_3SiCGeMe_3 + [Me_3SiCH_2PMe_3]Cl$$
$$\Vert$$
$$Me_3P$$

C. Complex salts and co-ordination complexes

In addition to the inorganic complex halides discussed in Chapter 2, it is reported that the trifluoromethylpentafluorogermanate ion is formed in aqueous solution by the reaction (Clark and Willis, 1962)—

$$F_3CGeI_3 + 5KF \rightarrow K_2[F_3CGeF_5] + 3KI$$

Ammonia and amine complexes of trimethyl- and triethyl-bromogermane ($R_3GeBr.NH_3$ and $R_3GeBr.NH_2R$) have been described, and several appear to be stable at room temperature (Kraus and Flood, 1932).

Phenanthrene co-ordination complexes of diphenyl- and di-n-butyl-dichlorogermane have been described. Their composition corresponds to $[R_2GeCl_2(phenan)]$; both melt with decomposition, but conductivity data have not been reported (Huber et al., 1966).

III. PSEUDOHALIDES

The chemistry of organometallic pseudohalides has been reviewed by Thayer and West (1967). Those groups so far reported in combination with germanium are cyanide (Anderson, 1951a, 1956a, 1961; Seyferth and Kahlen, 1960a, b), azide (Scherer and Schmidt, 1964; Thayer and West, 1964; Reichle, 1964), isocyanate (Anderson, 1949, 1951a, b, 1955, 1956a, 1961; Thayer and Strommen, 1966), isothiocyanate (Anderson, 1951a, 1953, 1956a, 1961; Rochow and Allred, 1955; Seyferth and Kahlen, 1960a) and fulminate (Beck and Schuierer, 1964). Germyl pseudohalides, H_3GeX are discussed in Chapter 2.

Although the chemistry of cyanide groups bonded to transition metals dates back to the beginning of co-ordination chemistry, with few exceptions (e.g., Et_3SnCN), non-transition metal cyanides have only been studied in recent years. Most germanium pseudohalide compounds have been prepared by exchange reactions on organogermanium-halides, -oxides, -sulphides or -hydrides using metal–pseudohalide salts or the free acids.

A. Germanium cyanides

Tetracyanogermane has been isolated from the reaction (Bither et al., 1958)—

$$4Me_3SiCN + GeCl_4 \rightarrow 4Me_3SiCl + Ge(CN)_4$$

The few properties examined suggest that it is polymeric, since it forms a buff powder insoluble in common organic solvents, and is rapidly hydrolysed by moist air, giving hydrogen cyanide. Organogermanium cyanides have been prepared by the action of silver cyanide on germanium halides or sulphides—

$$R_3GeI + AgCN \rightarrow R_3GeCN + AgI$$

$$(Pr^i_2GeS)_3 + 6AgCN \rightarrow 3Pr^i_2Ge(CN)_2 + 3Ag_2S$$

Although hydrogen cyanide reacts with organodigermoxanes in a similar way, diethylgermanium oxide as the cyclic trimer fails to react even with the addition of phosphorus pentoxide—

$$(R_3Ge)_2O + 2HCN \rightleftharpoons 2R_3GeCN + H_2O$$

A further preparative method is the action of a deficiency of mercuric cyanide on germanium dihydrides—

$$R_2GeH_2 + Hg(CN)_2 \rightarrow R_2Ge(H)CN + Hg + HCN$$

Triethylgermanium cyanide is described as a colourless liquid smelling of hydrogen cyanide.

The structure of tetracyanogermane is unknown, but it is difficult to see how it can associate other than by bridging groups, the probable result being either a trigonal bipyramidal or an octahedral grouping about germanium—

Its i.r. spectrum has not been reported but trimethylgermanium cyanide shows two bands in the region of the C≡N stretch (2197s and 2100w), possibly due to an equilibrium between cyanide and isocyanide, the higher frequency being ascribed to $\nu(GeC{\equiv}N)$. Evidence quoted in support of an equilibrium concentration of isocyanide is the reaction with sulphur, which gives the isothiocyanate—

$$Me_3GeCN \xrightleftharpoons{} Me_3GeNC \xrightarrow[170°C]{S} Me_3GeNCS$$

Triethylsilyl cyanide shows similar effects in the $\nu(C{\equiv}N)$ region; moreover the relative intensity of the two bands is temperature dependent, the proportion of isocyanide increasing with temperature. Alternative explanations that have been put forward are that the low-frequency band represents a C≡N vibration involving a thermally excited state or that it is due to an overtone or combination band. The crystal structure of trimethylcyanogermane shows only the cyanide form with no evidence for bridging by carbon–nitrogen groups (Schlemper and Britton, 1966). However this does not disprove the possibility of isomerism in solution, or as a liquid melt. The crystal structure of trimethyltin cyanide is quite different: it consists of pentaco-ordinate tin atoms in the form of infinite chains, each cyanide group being bonded to two tin atoms

$$C\equiv N \rightarrow \underset{\underset{Me}{\overset{|}{\overset{Me}{\underset{}{}}}}}{Sn} - C\equiv N \rightarrow \underset{\underset{Me}{\overset{|}{\overset{Me}{\underset{}{}}}}}{Sn} - C\equiv N \rightarrow$$

Few reactions have been reported on germanium cyanides. Hydrolysis, giving hydrogen cyanide, is evidently rapid and the cyanide group will undergo metathetical reactions with halide and other pseudohalide ions. The cyanide group can also function as an electron donor, and several complexes with Lewis acids have been isolated—

$$Me_3GeCN \begin{cases} \xrightarrow{Et_2OBF_3} Me_3GeCN \rightarrow BF_3 + Et_2O \\ \\ \xrightarrow[70^\circ C]{Fe(CO)_5} Me_3GeNC \rightarrow Fe(CO)_4 + CO \end{cases}$$

The boron trifluoride complex is a white crystalline solid that sublimes readily *in vacuo* at 40°C, and is rapidly hydrolysed in moist air. The iron tetracarbonyl complex, a yellow crystalline solid, also sublimes readily *in vacuo* and is decomposed rapidly by oxygen. This complex is considered to have the isocyanide rather than the cyanide structure mainly by analogy with the behaviour of iron pentacarbonyl towards organic nitriles and isocyanides, and comparative infrared data.

B. Azides

Several alkyl- and aryl-germanium azides of the types R_3GeN_3, $R_2Ge(N_3)_2$ and $RGe(N_3)_3$ have been reported. The high thermal stability of both silicon- and germanium-azides is noteworthy; triphenylgermanium azide decomposes about 375°C and does not appear to be shock sensitive. In common with other metal azides, both the symmetric and asymmetric –N–N–N stretching vibrations are i.r. active, and are intermediate in frequency between the values for similar silicon and tin compounds, the highest frequencies being observed with silicon (Me_3GeN_3, $\nu_{asym.} = 2102$, $\nu_{sym.} = 1290$; $Me_2Ge(N_3)_2$, $\nu_{asym.} = 2110$, $\nu_{sym.} = 1282$ cm^{-1}) (Thayer and Strommen, 1966). Trimethylgermanium azide is less rapidly hydrolysed than the silicon analogue—

$$2Me_3GeN_3 + H_2O \rightleftharpoons (Me_3Ge)_2O + 2HN_3$$

Triphenylgermyl azide decomposes above 375°C with the evolution of nitrogen, but the polymeric residue contains unidentified phenyl–nitrogen compounds. The azide group is cleaved by triphenylphosphine, liberating nitrogen, and forming triphenylgermylphosphinimine (Reichle, 1964)—

$$Ph_3GeN_3 + Ph_3P \rightarrow Ph_3GeN:PPh_3 + N_2$$

The u.v. spectra of trimethylgermyl- and trimethylsilyl-azide are almost identical, but as the $M—N_3$ bond becomes more ionic (as in Me_3SnN_3 and Me_3PbN_3) the absorption resembles that of the azide ion in aqueous solution.

Organogermanium azides, like the cyanides, behave as electron donors towards Lewis acids, and complexes with boron tribromide and tin(IV) chloride have been reported (Thayer and West, 1965)—

$$Ph_3GeN_3 . BBr_3 \qquad Ph_3GeN_3 . SnCl_4$$

These complexes however pose a structural problem since co-ordination may involve either the α- or the γ-nitrogen atom, e.g.—

$$\begin{matrix} Ph_3Ge \\ \\ Br_3\bar{B} \end{matrix} \!\! \diagdown \!\!\!\!\! \diagup \!\! \bar{N}—\overset{+}{N}\!\!\equiv\!\!N \qquad\qquad Ph_3GeN\!\!=\!\!\overset{+}{N}\!\!=\!\!N—\bar{B}Br_3$$

In valency-bond terms, two cannonical forms may be written for each structure, but no crystallographic evidence has yet been reported. It seems to be generally true that in co-ordination complexes of this type, the asymmetric –N–N–N stretching frequency shifts to higher frequency and the symmetric to lower frequency, relative to the unco-ordinated azide.

C. Isocyanates

Like the azides, three types have been reported, R_3GeNCO, $R_2Ge(NCO)_2$ and $RGe(NCO)_3$ where R = Me, Et, Pr^n, Pr^i and Bu^n. Their thermal stability is considerable and distillation may be carried out at atmospheric pressure. The rate of hydrolysis or alcoholysis increases with the number of isocyanato groups, e.g.—

$$Et_2Ge(NCO)_2 + 4ROH \rightarrow Et_2Ge(OR)_2 + 2H_2NCO_2R$$

Germanium tetraisocyanate, prepared by the reaction (Miller and Carlson, 1961)—

$$GeCl_4 + 4AgOCN \xrightarrow[78°C]{C_6H_6} Ge(NCO)_4 + 4AgCl$$

is of spectroscopic interest in relation to the silicon compound. The latter is tetrahedral with linear Si–N–C–O groups, suggestive of extensive π-bonding to silicon. The germanium compound shows only a single ^{13}C n.m.r. line and cannot therefore be a mixed compound, $Ge(NCO)_x(OCN)_{4-x}$.

Its i.r. and Raman spectra are compatible only with the isocyanate structure, but they also indicate that the structure deviates from tetrahedral symmetry.

D. Isothiocyanates

In addition to being formed by exchange reactions, organogermanium isothiocyanates have been prepared by the reaction of the cyanides with sulphur. Di-n-butylgermane, with a deficiency of mercuric thiocyanate, gives the hydride, $Bu^n_2Ge(H)NCS$. Only alkyl compounds R_3GeNCS, $R_2Ge(NCS)_2$ and $RGe(NCS)_3$ have been reported.

E. Fulminates

There is only one example of an organogermanium fulminate, Ph_3GeCNO.

It is obtained by an exchange reaction with silver fulminate as a colourless crystalline solid, insoluble in water and only slowly hydrolysed. It decomposes about 120°C.

8 | ORGANOGERMANIUM–METAL BONDED COMPOUNDS

This Chapter is devoted to the chemistry of compounds containing one or more bonds between germanium atoms (as in digermanes R_3GeGeR_3), together with compounds in which germanium is directly bonded to another metal or metalloid. Alkali metal- and magnesium halide-compounds, such as R_3GeLi and R_3GeMgX are included, although none has actually been isolated.

Bond-energy data on germanium–metal complexes is extremely limited. For all combinations of the Group IVB elements the silicon—silicon bond is the strongest (probably about 80 kcal. mole^{-1}), and the mixed silicon—germanium bond is probably somewhat stronger than the germanium—germanium bond in related compounds. Chemical reactivity of germanium–metal complexes varies widely; compounds with the more electropositive metals are frequently sensitive to oxygen, water and photolysis, whereas in many germanium–transition metal complexes the centre of reactivity is the transition metal.

I. CATENATED ORGANOGERMANES

Organo-derivatives of the group IVB elements form catenated compounds with all other members of the group. The range of compounds and their stability is, of course, greatest with carbon. Silicon, germanium and tin show much the same tendency to catenation, but a marked reduction is observed with lead. An excellent review of the whole field of catenated organo derivatives of the group IVB elements has been published (Gilman et al., 1966). In all of the catenated compounds of silicon, germanium, tin and lead, the metal is tetraco-ordinate, and the thermal stability decreases from silicon to lead. Differences in thermal stability between silicon and germanium are much less marked than for the heavier elements, but for all the metals the structure of the compound can markedly affect stability rather than the number of metal—metal bonds. This is strikingly illustrated with lead, where the remarkable red compound $(Ph_3Pb)_4Pb$, with four lead—lead bonds, is at least sufficiently stable to be isolated; the only other catenated lead compounds are of the type R_6Pb_2.

Catenated organogermanes may be sub-divided into four groups—

A. Digermanes, such as R_3GeGeR_3.
B. Cyclic organopolygermanes, such as cyclo-$(Ph_2Ge)_4$.
C. Linear and branched-chain species, such as $Me_3Ge(GeMe_2)_nGeMe_3$ and $(Ph_3Ge)_3GeH$.
D. Ill-defined high polymers.

There is a considerable increase in thermal stability and reduced chemical reactivity between the higher germanium hydrides discussed in Chapter 2 and the organo-di- and poly-germanes, and quite drastic chemical reactions can be employed in forming germanium—germanium bonds in organo-germanes.

A. Digermanes

Wurtz-type syntheses form the basis of the most general method for the synthesis of hexaorganodigermanes—

$$2R_3GeX + 2M \rightarrow R_3GeGeR_3 + 2MX$$

This reaction using an alkali metal or sodium–potassium alloy, either with or without solvent, has been used in the preparation of both hexa-alkyl- and hexa-aryl-digermanes. Yields are commonly of the order of 50%, and polygermanes are formed as by-products (Morgan and Drew, 1925; Bauer and Burschkies, 1934; Johnson and Nebergall, 1948; Shackelford *et al.*, 1963; Carrick and Glockling, 1966).

A related preparative method which basically involves coupling of a germyl-Grignard reagent with a germanium halide produces high yields of digermanes—

$$GeCl_4 + RMgX + Mg \rightarrow R_3GeMgX$$
$$R_3GeMgX + R_3GeX \rightarrow R_6Ge_2$$

The possible mechanism of this reaction is discussed later (p. 168); it has provided a good route to hexaphenyl- and the hexatolyl-digermanes, and has the virtue of being a one-stage synthesis (Glockling and Hooton, 1962b, 1966). Hexavinyldigermane has been obtained in 26% yield by this method (Seyferth, 1957). If mixed Grignard reagents are used, all combinations of digermanes are formed in a combined yield varying from 6 to 35% (Semlyen *et al.*, 1965)—

$$GeCl_4 + MeMgI + RMgBr + Mg \rightarrow Me_nR_{6-n}Ge_2$$

Mixed organolithium reagents react in the same way, but give ratios of products closer to those expected for a random distribution of alkyl groups. These reactions also provide chromatographic evidence of the formation of more highly catenated organogermanes.

In much the same way the alkylation of germanium tetrachloride by aluminium alkyls leads to the formation of digermanes in addition to the tetra-alkylgermane (Glockling and Light, 1967). The yield of digermane increases in the order $Me < Et < Bu^i$, and significant amounts of higher alkylpolygermanes, such as $Bu^i_8Ge_3$ are produced. It seems likely that germanium—germanium bond formation results from a reactive germanium–aluminium intermediate, analogous to the germyl-Grignard reagents—

$$R_3GeAlR_2 + R_3GeCl \rightarrow R_3GeGeR_3 + \tfrac{1}{2}(R_2AlCl)_2$$

The reaction of germanium tetrachloride with sodium and an organic halide has given $(p\text{-tolyl})_6Ge_2$ in reasonable yield, but using β-bromostyrene, only trace quantities of $(PhCH:CH)_6Ge_2$ are obtained (Birr and Kraeft, 1961)—

$$GeCl_4 + RCl + Na \rightarrow R_6Ge_2 + R_4Ge$$

Unsymmetrical digermanes are produced by condensation reactions on the pre-formed triorganogermyl–alkali metal derivative—

$$Ph_3GeNa + Et_3GeBr \xrightarrow{\ NH_3\ } Ph_3GeGeEt_3$$

$$(PhCH_2)_3GeLi + Et_3GeBr \xrightarrow{\ C_4H_{10}O_2\ } (PhCH_2)_3GeGeEt_3$$

The yields are low, largely owing to halogen–metal exchange and coupling to the symmetrical digermanes (Kraus and Sherman, 1933; Cross and Glockling, 1964). Triethyl- and tri-n-butylgermyl-potassium, which have been obtained in hexamethylphosphotriamide, $(Me_2N)_3PO$, solution have been used to prepare the unsymmetrical digermanes $Et_3GeGeMe_3$ and $Bu^n_3GeGeMe_3$ (Bulten and Noltes, 1966b, 1967, 1968). This appears to be a highly versatile reaction, which is markedly dependent on the order of addition of the reagents—

$$R_3GeK + Me_3GeCl \xrightarrow{\ (Me_2N)_3PO\ } R_3GeGeMe_3$$

If trimethylchlorogermane is added to triethylgermylpotassium a 55% yield of the unsymmetrical digermane is obtained, whereas reversing the addition increases the yield to 83%. This suggests that the digermane is cleaved by Et_3GeK in a rapid reaction, and it has been established that trimethylgermyl-lithium will cleave hexaethyldigermane—

$$Me_3GeLi + Et_6Ge_2 \xrightleftharpoons{(Me_2N)_3PO} Et_3GeGeMe_3 + Et_3GeLi$$

Similarly triethyltrimethyldigermane is stable towards disproportionation in hexamethylphosphotriamide solution, but the reaction is strongly catalysed by trialkylgermylpotassium, or other nucleophiles—

$$2Et_3GeGeMe_3 \xrightleftharpoons{R_3GeK, (Me_2N)_3PO} Et_6Ge_2 + Me_6Ge_2$$

An approximate equilibrium constant of 0·12 has been evaluated for the forward reaction.

Symmetrical tetraphenyldiethyldigermane has been obtained by the following two routes—

$$GeI_2 \xrightarrow{PhLi} \text{yellow polymer} \xrightarrow{Br_2} Ph_2BrGeGeBrPh_2$$
$$\downarrow EtMgBr$$
$$Ph_2EtGeGeEtPh_2$$
$$\nearrow EtBr$$
$$Ph_2GeH_2 \xrightarrow{BuLi} Ph_2(Li)GeGe(Li)Ph_2$$

It has also been reported that di-n-butylmercury and germanium di-iodide react to give the tetrabutyldi-iododigermane although this compound has not been thoroughly characterized (Jacobs, 1954).

$$GeI_2 + Bu^n_2Hg \rightarrow Bu^n_2(I)GeGe(I)Bu^n_2 + Bu^nHgI$$

Mild reduction of phenyltribromogermane and diphenyldibromogermane results in coupling to the digermane (Metlesics and Zeiss, 1960)—

$$PhGeBr_3 + Li/Hg \xrightarrow{Et_2O} Ph(Br)_2GeGe(Br)_2Ph$$
$$Ph_2GeBr_2 + Li/Hg \xrightarrow{Et_2O} Ph_2BrGeGeBrPh_2$$

Various organogermyl–metal complexes decompose thermally or by irradiation to give the hexaorganodigermane.

B. Cyclic organopolygermanes

Application of the Wurtz reaction to diphenyldihalogermanes, Ph_2GeX_2, produces both cyclic- and high-molecular-weight linear or branched-chain polymers. The proportion of 4-, 5- and 6-membered compounds depends on the particular alkali metal, the dilution and the solvent employed. Six-membered rings are the largest so far isolated (Kraus and Brown, 1930b; Neumann and Kuehlein, 1965)—

The cyclic tetragermane has been prepared from diphenylgermane and diethylmercury via a polymer that is decomposed by heat or u.v. light—

$$Ph_2GeH_2 + Et_2Hg \rightarrow C_2H_6 + [Ph_2GeHg]_n \rightarrow (Ph_2Ge)_4 + Hg + (Ph_2Ge)_n$$

Reduction of dimethyldichlorogermane by lithium leads to a cyclic hexamer together with polymeric products (Nefedov et al., 1962)—

$$Me_2GeCl_2 \xrightarrow{Li, THF} (Me_2Ge)_6$$

C. Linear and branched-chain polygermanes

It seems likely that most Grignard and similar reagents, in their reactions with germanium tetrachloride produce small proportions of tri-, tetra- and higher-polygermanes (Quane and Hunt, 1968). The Wurtz reaction applied to trimethylbromogermane yields, in addition to hexamethyldigermane, more complex germanes resulting from the cleavage of either a germanium—carbon or carbon—hydrogen bond, followed by further coupling processes—

$$Me_3GeBr \xrightarrow{K} Me_6Ge_2 + Me_3GeCH_2Ge(Me)_2GeMe_3 + (Me_3Ge)_4Ge$$

Trimethylaluminium and germanium tetrachloride give low yields of mixed methylpolygermanes, and, by the use of high-resolution mass spectrometry, at least five have been unambiguously identified (Me_6Ge_2, Me_8Ge_3, $Me_{10}Ge_4$, $Me_{12}Ge_5$, $Me_{14}Ge_6$). A further feature of interest in these reactions is the formation of polygermanes containing longer alkyl groups. For example, trimethylaluminium produces Me_7Ge_3Et, $Me_6Ge_3Et_2$

and Me_9Ge_4Et, whereas triethylaluminium and germanium tetrachloride give low-yield products containing both butyl and hexyl groups, (e.g., Et_7Ge_3Bu and $Et_5Ge_2C_6H_{13}$). Tri-isobutylaluminium and germanium tetrachloride form, in addition to tetraisobutylgermane, isolable amounts of the di- and tri-germanes $Bu^i_6Ge_2$ and $Bu^i_8Ge_3$.

The mechanism of formation of polygermanes and the lengthening of alkyl chains is rather obscure. Reactive Ge–Al intermediates are almost certainly involved, and it is likely that reduction of germanium(IV) to germanium(II) occurs either as a primary reaction of germanium tetrachloride or at an intermediate stage (such as Me_2GeCl_2) in the alkylation process. These two requirements are sufficient to account for the formation of linear and branched-chain methylpolygermanes, in which Me_3Ge groups are chain terminating. The formation of $GeCH_2Ge$ groups, which is a further side reaction, and the growth of alkyl chains is most readily explained if chlorination by aluminium chloride and further condensation reactions are invoked. Alternatively an intermediate like $>GeCH_2Al(H)-$ may be produced—

$$>Ge-CH_3 \xrightarrow{AlCl_3} >Ge-CH_2Cl \xrightarrow{Me_3GeAlCl_2} >GeCH_2GeMe_3$$

Processes of the following types illustrate the formation of linear methyl-polygermanes—

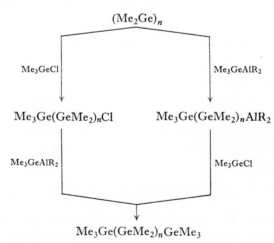

$$Me_3Ge(GeMe_2)_n GeMe_3$$

Germanium di-iodide undergoes interesting reactions with organo-metallic compounds. For example, with triphenylgermyl-lithium it gives the branched chain Ge_4 compound (Glockling and Hooton, 1963).

$$3Ph_3GeLi + GeI_2 \longrightarrow (Ph_3Ge)_3GeLi \begin{array}{c} \overset{H_2O}{\longrightarrow} (Ph_3Ge)_3GeH \\ \underset{MeI}{\searrow} \\ (Ph_3Ge)_3GeMe \end{array}$$

This reaction, which also produces highly coloured by products, probably proceeds via a germanium(II) intermediate, $(Ph_3Ge)_2Ge$.

Germanium di-iodide and excess trimethylaluminium react smoothly at room temperature in pentane solution, forming a range of methylpoly-germanes. In the course of the reaction germanium(II) is oxidized to ger-manium(IV), hence a "reduced complex" (such as the halogen-bridged species, Me_2GeAlI_3) must be present, and subsequent oxidative hydrolysis results in the formation of tetrameric dimethylgermanium oxide. Methyl-polygermanes actually isolated include $Me_{10}Ge_4$, $Me_{12}Ge_5$, $Me_{14}Ge_6$ and $Me_{16}Ge_7$, and species up to $Me_{22}Ge_{10}$ have been identified by mass spectro-metry. High-molecular-weight telomers are present, together with products resulting from the growth of hydrocarbon chains (Me_9Ge_4Et, Me_9Ge_4Pr).

The identification of isomeric alkylpolygermanes is difficult since de-gradation by, for example, bromine is liable to cleave germanium—carbon bonds and this would invalidate structural conclusions. A combination of chromatography, mass spectrometry and high-resolution p.m.r. at 220 Mc/s provides the complete answer for methylpolygermanes, since the proton chemical shifts of Me_3Ge, Me_2Ge and $MeGe$ groups are sufficiently different for their unambiguous recognition. In this way it has been shown that all degrees of chain branching occur as in the following examples (Glockling et al., 1968)—

$$\begin{array}{cc}
(Me_3Ge)_2GeMe & (Me_3Ge)_2GeMe \\
| & | \\
Me_3GeGeMe_2 & (Me_3Ge)_2GeMe \\
\\
Me_3Ge(GeMe_2)_3GeMe_3 & (Me_3Ge)_3GeMe
\end{array}$$

Similar products are produced by the reaction of methyl-lithium or methylmagnesium iodide with trichlorogermane in ether solution (Nefedov and Kolesnikov, 1966). Trichlorogermane may well behave as germanium dichloride in this reaction and some 5 % of each of the following compounds has been isolated: Me_6Ge_2, Me_8Ge_3, $Me_{10}Ge_4$, $Me_{16}Ge_7$, together with a mixture of liquid and solid higher telomers.

High-molecular-weight, presumably linear, polymers are formed by the action of lithium on dimethyldichlorogermane, but there is no information

6

on the nature of the end-groups. The i.r. spectrum of this polydimethyl-germanium corresponds closely to that obtained from the reaction between germanium di-iodide and trimethylaluminium (Nefedov *et al.*, 1962)—

$$Me_2GeCl_2 \xrightarrow{\text{Li, THF}} (Me_2Ge)_n + (Me_2Ge)_6$$

Octaphenyltrigermane, a colourless crystalline solid, has been obtained from the reaction between triphenylgermylsodium and diphenyldichlorogermane (Kraus and Brown, 1930a)—

$$2Ph_3GeNa + Ph_2GeCl_2 \rightarrow Ph_3GeGePh_2GePh_3$$

Use has also been made of cyclic diphenylgermanium compounds as a means of forming unbranched polygermanes—

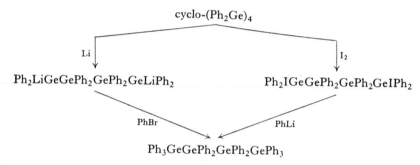

Linear ethylpolygermanes have been prepared from triethylgermyl-lithium (Bulten and Noltes, 1967)—

$$Et_3GeGe(Cl)Et_2 \xrightarrow{Et_3GeK} Et_8Ge_3$$

$$Et_2(Cl)GeGe(Cl)Et_2 \xrightarrow{Et_3GeK} Et_{10}Ge_4$$

D. Ill defined high polymers

Various reactions have been reported as yielding polymeric organoger-manes which necessarily require extensive metal–metal bonding. When diphenyldichlorogermane is reduced by sodium in boiling xylene, the colourless tetramer $(Ph_2Ge)_4$ is produced in 10–20% yield, together with much yellow resinous material having a mean molecular weight of about 900 and containing 30–32% of germanium (Ph_2Ge requires 32%). Polymeric "diphenylgermanium" of somewhat similar appearance is produced in the reaction between germanium di-iodide and excess phenyl-lithium as a yellow amorphous powder having a mean molecular weight in the range 800–1700. Controlled degradative bromination provides evidence that this yellow polymer contains Ph_3Ge, Ph_2Ge, $PhGe$ and Ge units, giving an overall composition intermediate between Ph_2Ge and $PhGe$. This rather surprising result has been interpreted by assuming that phenylation of germanium di-iodide is in competition with halogen–metal exchange, so that species such as LiGeI and PhGeLi form reactive intermediates in the poly-merization processes. Since both triphenylgermane and tetraphenylgermane are finally isolated, the following reactions must also occur—

$$Ph_2Ge + PhLi \longrightarrow Ph_3GeLi \begin{array}{c} \overset{H_2O}{\nearrow} Ph_3GeH \\ \underset{\searrow}{\overset{PhI}{\longleftarrow}} \\ Ph_4Ge \end{array}$$

$$GeI_2 + PhLi \longrightarrow LiGeI + PhGeI + PhGeLi + PhI$$

This yellow polymer is apparently free from halogen and oxygen, but is slowly oxidized by air. Triphenylaluminium phenylates germanium di-iodide, giving a similar yellow amorphous polymer containing oxygen, presumably derived from the hydrolysis of residual germanium–iodine groups (Glockling and Hooton, 1963).

Similar intractable polymers are produced by the action of alkali metals on phenyltrichlorogermane, but even after long reaction times these contain both chlorine (1–5 %) and oxygen (5–10 %) (Schwarz and Schmeisser, 1936; Metlesics and Zeiss, 1960). There are several reports of polymeric organo-germanes that approximate in composition to RGe, but which could con-ceivably be organogermanium oxides. Thus isopropylmagnesium bromide and germanium tetrabromide are said to give a polymer $(Pr^iGe)_n$, although later work on isopropylgermanes failed to confirm this report (Anderson, 1953; Carrick and Glockling, 1966). n-Propyl-lithium and germanium tetrachloride apparently give a propylgermanium polymer, Pr^nGe, as a colourless viscous liquid volatile *in vacuo*. It is difficult to reconcile this formulation with its volatility. Decomposition of hexaethyldigermane by aluminium halides at 150°–200°C produces tetraethylgermane and a dark tarry product of composition EtGe (Vyazankin *et al.*, 1964b), whereas pyrolysis of 1,2-dichlorotetraethyldigermane at 200°C yields a polymeric diethylgermanium (Bulten and Noltes, 1966a)—

$$Et_2(Cl)GeGe(Cl)Et_2 \xrightarrow{200°C} Et_2GeCl_2 + (Et_2Ge)_n$$

The spectroscopic properties of polygermanes are of particular interest. Cyclic-, and to a lesser degree linear-, polysilanes and -polygermanes form paramagnetic anions with sodium–potassium alloy in tetrahydrofuran and similar solvents. These charge-transfer complexes are unstable above about −50°C, when they decompose to unidentified diamagnetic products. The e.s.r. spectrum of the silane, cyclo-$(Me_2Si)_6$, has been analysed, showing that the added electron is equally delocalized over all six silicon atoms and therefore contacts equally all of the protons. This means that the electron must occupy a π (or δ) type orbital derived from the $3d$ orbitals of the six silicon atoms, and the ion radical need not be planar for effective delocalization over the ring. However it is necessary to postulate interconversion of ring conformations in order to make all the groups equivalent on a time-average basis (Husk and West, 1965; West and Carberry, 1967).

Digermanes absorb strongly in the near u.v. (Ph_6Ge_2, $\lambda_{max} = 239$ nm, $\epsilon = 3\cdot04 \times 10^4$; $Pr^i_6Ge_2$, $\lambda_{max} = 210$ nm, $\epsilon = 6\cdot46 \times 10^{-3}$). From comparative studies on related silicon, tin and lead compounds it is apparent that this long-wavelength absorption is due to the metal–metal bond acting as a chromophore (Hague and Prince, 1964; Drenth et al., 1964). In phenyl-polysilanes of the type $Ph(Me_2Si)_nPh$, both λ_{max} and ϵ increase as n increases, although the cyclic silane $Me_{12}Si_6$ has an absorption band 231 nm (Gilman et al., 1966).

E. Reactions of organopolygermanes

Most of the information available relates to digermanes, although some reactions have been studied on cyclic polygermanes.

Surprisingly little information is available on the thermal stability of even the simple hexa-alkyl- and hexa-aryl-digermanes. Hexaphenyldigermane melts in air at 352°–354°C without decomposition, and hexaethyldigermane can be distilled in air at 265°C. Even the higher methylpolygermanes will survive gas–liquid chromatography at 230°C, but pyrolysis of $(Me_2Ge)_n$ at 350°–400°C results in degradation to the cyclic hexamer $(Me_2Ge)_6$. Both octaphenyltrigermane and hexacyclohexyldigermane decompose at their melting points (238° and 316°C, respectively). By comparison hexaethylditin decomposes at 275°C—

$$Et_6Sn_2 \rightarrow Et_4Sn + Sn + 2C_2H_5^{\cdot}$$

In no case have the products of thermal decomposition of digermanes been studied and there is no evidence for homolytic fission of the germanium—germanium bond although, under electron impact, organodigermanes show metastable-supported processes for the elimination of R_3Ge radicals (Glockling and Light, 1968), e.g.—

$$Ph_6Ge_2^{+\cdot} \rightarrow Ph_3Ge^+ + Ph_3Ge^{\cdot}$$

A preliminary account has been given of the pyrolysis of methylpolygermanes in the presence of ethylene (Nefedov et al., 1965b). This thermal cracking results in the formation of reactive Me_2Ge monomers, which combine with ethylene to give a variety of products—

$$(Me_2Ge)_n \xrightarrow[350°C]{C_2H_4} Me_2Ge\langle\rangle GeMe_2 + Me_2Ge\langle\rangle + [Me_2Ge(CH_2CH_2)_n]_m$$

Similarly the formation of cyclic or polymeric $(R_2Ge)_n$ compounds by the reductive methods described earlier may proceed through monomeric intermediates showing carbenoid character. Several of these reactions have been carried out in the presence of an alkene when low yields of cyclic germanes are produced—

Apart from the amorphous yellow polymeric materials, all organopoly-germanes are stable in air. They are degraded by hot oxidizing acids in the same way as tetraorganogermanes giving, finally, germanium dioxide. Fuming nitric- and sulphuric-acids, usually with the addition of ammonium persulphate, have been used for their analytical conversion to germanium dioxide. Several of the coloured amorphous phenylgermanium polymers, by contrast, take fire in nitric acid. Silver nitrate in boiling ethanol is said to oxidize hexaphenyldigermane (Morgan and Drew, 1925).

Bromination cleavage of germanium—germanium bonds occurs under quite mild conditions. Most studies have been made on hexaphenyldiger-mane, where further cleavage of a germanium—phenyl bond can occur (Kraus and Foster, 1927; Glockling and Hooton, 1963)—

Hexavinyldigermane is cleaved by iodine in hot chloroform, whereas hexaphenyldigermane reacts only slowly, even in refluxing xylene. Although cyclo-$Ph_{12}Ge_6$ and cyclo-$Ph_{10}Ge_5$ are unattacked by iodine in benzene, the 4-membered ring compound cyclo-Ph_8Ge_4 is more reactive—

$$\begin{array}{c} \text{Ph}_2\text{Ge—GePh}_2 \\ |\quad\quad| \\ \text{Ph}_2\text{Ge—GePh}_2 \end{array} \xrightarrow{\text{I}_2} \text{Ph}_2\text{IGeGePh}_2\text{GePh}_2\text{GeIPh}_2$$

Hexaethyldigermane reacts with isopropyl bromide in the presence of aluminium chloride, partly by cleavage of a germanium—ethyl bond, whereas with aluminium chloride or bromide alone at 150°–200°C, further decomposition occurs, presumably with loss of hydrocarbons (Vyazankin et al., 1964b)—

$$\text{Et}_6\text{Ge}_2 \begin{cases} \xrightarrow{\text{AlCl}_3,\ \text{Pr}^i\text{Br}} \text{Et}_3\text{GeGeClEt}_2 + \text{Et}_4\text{Ge} + \text{Et}_3\text{GeBr} + \text{tar} \\ \\ \xrightarrow[200°\text{C}]{\text{AlCl}_3} \text{Et}_4\text{Ge} + (\text{EtGe})_n \end{cases}$$

Redistribution reactions between hexa-alkyldigermanes and germanium tetrachloride proceed under quite mild conditions: hexamethyldigermane redistributes when refluxed with germanium tetrachloride (Rijkens et al., 1966; Glockling and Light, unpublished work)—

$$\text{Me}_6\text{Ge}_2 \xrightarrow{\text{GeCl}_4} \text{Me}_5\text{Ge}_2\text{Cl}$$

$$\text{Et}_6\text{Ge}_2 \xrightarrow[200°\text{C}]{\text{GeCl}_4} \text{Et}_5\text{Ge}_2\text{Cl} + \text{Et}_2(\text{Cl})\text{GeGe}(\text{Cl})\text{Et}_2$$

Tetra-alkylgermanes are unreactive under these conditions without a catalyst. Halogen acid cleavage of germanium—germanium bonds has only been reported for methylpolygermanes (Nefedov and Kolesnikov, 1966)—

$$\text{Me}_3\text{Ge}(\text{Me}_2\text{Ge})_n\text{GeMe}_3 \xrightarrow[250°\text{C}]{\text{HCl}} \text{Me}_2\text{GeCl}_2 + \text{Me}_2\text{Ge}(\text{H})\text{Cl}$$

Cleavage of germanium—germanium bonds by alkali metals is discussed later (p. 161).

II. SILICON, TIN AND LEAD COMPOUNDS

A. Germanium–silicon compounds

Several germanium–silicon compounds have been prepared either by a "mixed" Wurtz reaction or by reactions involving elimination of an alkali metal halide (Milligan and Kraus, 1950; Gilman and Gerow, 1956b; Shackelford et al., 1963; Cross and Glockling, 1964)—

$$\text{Et}_3\text{GeBr} + \text{Et}_3\text{SiBr} \xrightarrow{\text{Na, THF}} \text{Et}_3\text{GeSiEt}_3 + \text{Et}_6\text{Si}_2 + \text{Et}_6\text{Ge}_2$$

$$\text{Ph}_3\text{SiK} + \text{Ph}_3\text{GeBr} \longrightarrow \text{Ph}_3\text{GeSiPh}_3$$

$$(\text{PhCH}_2)_3\text{GeLi} + \text{Me}_3\text{SiCl} \longrightarrow (\text{PhCH}_2)_3\text{GeSiMe}_3$$

Triethylgermyl-lithium behaves like the aryl derivatives towards both chlorosilanes and silicon hydrides, the former being the more reactive (Vyazankin *et al.*, 1966a)—

$$(Et_3Ge)_2SiPh_2 \xleftarrow{\text{Ph}_2\text{SiCl}_2} Et_3GeLi \begin{cases} \xrightarrow{\text{HSiCl}_3} (Et_3Ge)_3SiH \\ \xrightarrow{\text{Et}_3\text{SiH}} Et_3GeSiEt_3 \end{cases}$$

Trimethylsilylgermane and trimethylgermylsilane have been reported; the latter is readily hydrolysed in alkaline solution with liberation of hydrogen (Amberger and Muehlhofer, 1968).

In both methods extensive formation of the symmetrical germanium–germanium and silicon–silicon compounds occurs. The extent of halogen–metal exchange, which is responsible for formation of symmetrical products in the second type of reaction is dependent on the order of addition of the reactants and on the halide employed. Iodides are more prone to halogen–metal exchange than chlorides, and it appears that halogermanes are less susceptible to halogen–metal exchange than halosilanes. A further synthetic method is the pyrolysis, or u.v.-induced decomposition, of mercury derivatives (Vyazankin *et al.*, 1964a)—

$$Et_3GeHgSiEt_3 \xrightarrow{hv} Et_3GeSiEt_3 + Hg$$

Germanium–silicon compounds are highly stable and are of low reactivity. Thus $Ph_3GeSiPh_3$ melts at 357°C and volatilizes at 500°C with some decomposition. The germanium—silicon bond is cleaved by sodium–potassium alloy in ether; in this respect it resembles hexaphenyldisilane rather than hexaphenyldigermane—

$$Ph_3GeSiPh_3 \xrightarrow{\text{Na/K}} Ph_3GeK + Ph_3SiK$$
$$\downarrow{CO_2}$$
$$Ph_3GeCO_2H + Ph_3SiCO_2H$$

Triphenylgermylsodium and trichlorosilane in liquid ammonia give the tris compound $(Ph_3Ge)_3SiH$, which in several of its reactions shows the low reactivity of the germanium—silicon bond—

$$(Ph_3Ge)_3SiBr \xleftarrow{\text{Br}_2} (Ph_3Ge)_3SiH \xrightarrow{\text{Li, EtNH}_2} (Ph_3Ge)_3SiLi$$

The bromide $(Ph_3Ge)_3SiBr$, when boiled with aqueous sodium hydroxide, decomposes giving hexaphenyldigermane quantitatively. By contrast, tri-phenylgermyl-lithium and silicon tetrachloride give only amorphous material corresponding to the composition $(Ph_3Ge)_4Si_2$. Similarly when tribenzylgermyl-lithium is added to trichlorosilane only low-melting polymeric material is produced.

B. Germanium–tin compounds

A somewhat wider range of compounds having germanium—tin bonds are known. Two basically different preparative methods have been employed; the first using a metal halide, can be applied in various ways (Kraus and Foster, 1927; Gilman and Gerow, 1957a; Willemsen and Van der Kerk, 1964; Gilman and Cartledge, 1966; Wiberg et $al.$, 1967)—

$$Ph_3GeK \xrightarrow{Ph_3SnCl} Ph_3GeSnPh_3$$

$$Ph_3GeLi \xrightarrow{SnCl_2} (Ph_3Ge)_3SnLi \xrightarrow{Ph_3GeCl} (Ph_3Ge)_4Sn$$

$$SnCl_4$$

$$Ph_3SnLi \Big\langle \xrightarrow{GeCl_4} (Ph_3Sn)_4Ge$$

$$\xrightarrow{SnCl_2} (Ph_3Sn)_3SnLi \xrightarrow{Ph_3GeCl} (Ph_3Sn)_3SnGePh_3$$

The tetrakis complex $(Ph_3Sn)_4Ge$ is remarkably stable, decomposing at 342°C. This and related compounds are of course highly symmetrical, and the stability seems to decrease with the mass of the central metal atom.

An extremely versatile method of establishing germanium—tin bonds makes use of the reactivity of either tin–nitrogen or germanium–nitrogen compounds towards protic reagents. It has the advantage of avoiding halogen–metal exchange, which can make the isolation of reaction products quite difficult—

$$Bu^n_3GeNEt_2 + Ph_3SnH \rightarrow Bu^n_3GeSnPh_3 + Et_2NH$$

This reaction, which is slow in hot butyronitrile, may be extended in a variety of ways, as in the following example (Neumann et $al.$, 1966; Creemers and Noltes, 1967).—

$$Bu^n_3GeNMe_2 + Bu^n_2SnH_2 \longrightarrow Bu^n_3GeSn(Bu^n)_2H$$

$$\Big\downarrow Et_2NH$$

$$Bu^n_3Ge(SnBu^n_2)_2GeBu^n_3 + H_2$$

Organotin hydrides react with triethylgermyl-lithium in the same way as silanes—

$$Et_3GeLi \xrightarrow{Et_3SnH} Et_3GeSnEt_3$$

Despite the number of germanium–tin compounds prepared, very few reactions have been studied. The germanium—tin bond is cleaved selectively by n-butyl-lithium and by hot acetic acid. The acetoxy compound $(AcO)_3GeSn(OAc)_3$ is stable to 360°C; some of its reactions are mentioned in Chapter 2—

$$Ph_3GeSnPh_3 \xrightarrow{BuLi} Ph_3GeLi + Ph_3SnBu$$

$$\downarrow AcOH \qquad\qquad\qquad \downarrow CO_2$$

$$Ph_3GeSn(OAc)_3 \qquad\qquad Ph_3GeCO_2H$$

$$\downarrow AcOH$$

$$(AcO)_3GeSn(OAc)_3$$

C. Germanium–lead compounds

The first germanium–lead compounds to be prepared are also the most complex—

$$Ph_3PbLi + GeCl_4 \rightarrow (Ph_3Pb)_4Ge$$

$$Ph_3GeLi + PbCl_2 \rightarrow [(Ph_3Ge)_2Pb] \rightarrow (Ph_3Ge)_4Pb + Pb$$

Both of these tetrakis derivatives are yellow, and $(Ph_3Pb)_4Ge$ decomposes at 210°C (Willemsens and Van der Kerk, 1964). Simpler germanium–lead compounds have been obtained by making use of the high reactivity of lead—nitrogen bonds in reactions analogous to those used for tin—germanium bond formation (Neumann and Kuehlein, 1966)—

$$Ph_3PbNR_2 + Ph_3GeH \rightarrow Ph_3PbGePh_3 + R_2NH$$

$$2Ph_3PbNR_2 + Ph_2GeH_2 \rightarrow (Ph_3Pb)_2GePh_2 + 2R_2NH$$

III. ALKALI-METAL COMPOUNDS

Compounds of the type R_3GeM (M = alkali metal) are of considerable use in the preparation of unsymmetrical organogermanes and many other derivatives. They have invariably been formed and used without isolation or purification, and there is no information on whether they are associated, or the extent to which they are solvated. The only structural information is derived from the formation and reactions of asymmetric lithiogermanes, $R_1R_2R_3GeLi$, where it has been shown that, in ether solution, racemization does not occur. This high optical stability implies a close ion-pair association between the solvated lithium cation and the $R_1R_2R_3Ge$ anion (p. 65). In a solvent of higher dielectric constant, such as ammonia, racemization is more likely to occur.

6*

The formation and stability of triorganogermyl–alkali metal compounds is profoundly influenced by the solvent, and solutions vary in colour from yellow to brown. Triethylgermyl potassium was first obtained in an almost quantitative reaction by cleavage of the digermane using potassium in ethylamine solution (Kraus and Flood, 1932); it is also formed in liquid ammonia solution, but it is then ammonolysed—

$$Et_3GeGeEt_3 + 2K \xrightarrow{EtNH_2} 2Et_3GeK$$

Triethylgermyl-lithium has been prepared in a similar reaction, but it is even more readily ammonolysed in liquid ammonia and fairly rapidly in ethylamine solution—

$$Et_3GeGeEt_3 + 2Li \xrightarrow{NH_3} 2Et_3GeLi \xrightarrow{NH_3} Et_3GeH + LiNH_2$$

Differences in the rate of solvolysis of the potassium and lithium salts probably reflect variations in the extent of ion-pairing, and the fact that potassium amide is freely soluble in liquid ammonia, whereas lithium amide is only sparingly soluble. Solvolysis therefore imposes a limitation to the usefulness of trialkylgermyl–alkali metal derivatives formed in this way, and of course many of the reagents (e.g., acid chlorides, metal halides) that might be used in reactions are also subject to rapid reaction with ammonia or primary amines. Nevertheless, some reactions have been carried out on triethylgermylpotassium—

$$Et_3GeK + EtBr \xrightarrow{NH_3} Et_4Ge + KBr$$

There have been no examples of trialkylgermyl derivatives of this type being obtained in useful concentrations by the cleavage of an alkyl group from a tetra-alkylgermane. Tetraethyl- and tetra-n-octyl-germane give coloured solutions with sodium–potassium alloy in 1,2-dimethoxyethane, but no derivatives have been isolated (Gilman et al., 1959). Two methods of preparing trialkylgermyl metal derivatives have been described recently; both proceed in high yield and are likely to be applicable to a wide range of alkyl and aryl groups. Bistrialkylgermylmercury compounds react quantitatively with lithium in tetrahydrofuran over 1 hr at 20°C (Vyazankin et al., 1967; Chambers and Glockling, 1968), or more slowly (3 days at 20°C) in benzene (Vyazankin et al., 1966b)—

$$Et_3GeBr + Hg/Na \xrightarrow{cyclohexane} (Et_3Ge)_2Hg$$

$$(Et_3Ge)_2Hg + 2Li \xrightarrow{THF} 2Et_3GeLi + Hg$$

The thallium derivative, $(Et_3Ge)_3Tl$, is cleaved by lithium in the same way.

The other method of obtaining R_3GeM compounds is the cleavage of a germanium—halogen or germanium—germanium bond by an alkali metal in hexamethylphosphotriamide, $(Me_2N)_3PO$. This solvent (Normant, 1967), which dissolves alkali metals forming soluble charge-transfer complexes, produces reddish-brown solutions of R_3GeM compounds, which are evidently stable over at least 3 weeks at $0°C$. Like the germylmercury method, this gives essentially quantitative yields, in reactions requiring some 4 hr at $20°C$ (Bulten and Noltes, 1966b, 1967)—

$$(Me_2N)_3PO \xrightarrow{Na} Na[(Me_2N)_3PO]^{-\cdot} \xrightarrow{Na} Na_2[(Me_2N)_3PO]^{2-}$$

$$Et_3GeGeEt_3 + 2K \xrightarrow{(Me_2N)_3PO} 2Et_3GeK$$

$$Bu^n_3GeCl + 2K \xrightarrow{(Me_2N)_3PO} Bu^n_3GeK + KCl$$

Triarylgermyl-alkali metal compounds, especially Ph_3GeLi, have been more extensively used as synthetic reagents. They can be prepared by cleavage of germanium—aryl, germanium—hydrogen, germanium—germanium or germanium—halogen bonds. The first reported compound, triphenylgermylsodium, was obtained by cleavage of hexaphenyldigermane with sodium in liquid ammonia (Kraus and Foster, 1927)—

$$Ph_6Ge_2 + 2Na \xrightarrow{NH_3} 2Ph_3GeNa$$

When prepared from tetraphenylgermane and sodium in liquid ammonia, phenylsodium, an immediate product from the cleavage reaction, is rapidly solvolysed to benzene and sodamide. The blue colour of sodium in liquid ammonia changes to orange–brown as the reaction proceeds and, with an excess of sodium, a red solution is formed, which is said to contain the disodio-derivative, Ph_2GeNa_2, but evidence in support of this is lacking. Triphenylgermylpotassium is formed by the cleavage of hexaphenyldigermane using sodium–potassium alloy in diethyl- or dibutyl-ether, in which it is insoluble (Gilman and Gerow, 1955c, 1957b). The most convenient and uncomplicated method for preparing triphenylgermyl-lithium is the action of butyl-lithium on the hydride (Gilman and Gerow, 1956a)—

$$Ph_3GeH + Bu^nLi \xrightarrow{Et_2O} Ph_3GeLi + C_4H_{10}$$

The other methods require the use of tetrahydrofuran or 1,2-dimethoxy-ethane as solvent, and complications due to attack of R_3GeLi on the solvent occur to some extent (Gilman and Gerow, 1955a, 1955b; George et al., 1960). The conclusion of Tamborski et al. (1962) that cleavage of hexaphenyldigermane by lithium is the most satisfactory preparative method is doubtful, since this reaction is often difficult to start, and complete separation of a monoglyme solution from finely divided lithium is an acute complication in some cases—

$$Ph_3GeGePh_3 + 2Li \xrightarrow{C_4H_{10}O_2} 2Ph_3GeLi$$

Cleavage of tetraphenylgermane by lithium in tetrahydrofuran or mono-glyme involves the formation of phenyl-lithium, which must rapidly react with the solvent, since benzoic acid is not obtained by subsequent carbona-tion—

$$Ph_4Ge + 2Li \xrightarrow{C_4H_{10}O_2} Ph_3GeLi + PhLi$$

Triphenylchlorogermane has been studied, but it seems to react rather slowly and incompletely; moreover conditions must be such that coupling between Ph_3GeLi and Ph_3GeCl to the digermane is minimized.

Solvent cleavage by $(aryl)_3GeLi$ compounds has not been extensively studied. The reactions between lithium and tetraphenylgermane or hexa-phenyldigermane are usually carried out between 0° and 20°C using a slurry of the arylgermane and lithium shot with a minimal amount of solvent. After a period, varying from a few minutes to several hours, bright yellow or orange areas appear in the slurry, and at this stage more solvent is added, usually with external cooling since the reactions are exothermic. Refluxing tetrahydrofuran solutions of triphenylgermyl-lithium produces the solvent cleavage product in 10% yield (Gilman and Zuech, 1961), whereas in 1,2-dimethoxyethane low yields of methyltriphenylgermane have been isolated.

The reactivity of R_3MLi compounds has been compared by the rate at which they cleave tetrahydrofuran, giving the order: $Bu_3SnLi > Ph_3SnLi$, $Ph_3GeLi > Ph_3SiLi$ (Gilman et al., 1965)—

$$Ph_3GeLi \begin{cases} \xrightarrow[\text{2. } H_2O]{\text{1. } C_4H_8O} Ph_3Ge(CH_2)_4OH \\ \xrightarrow[\text{2. } H_2O]{\text{1. } C_4H_{10}O_2} Ph_3GeMe + MeOCH_2CH_2OH \end{cases}$$

Further information on the stability of triphenylgermyl-lithium has been obtained by treating with n-octadecyl bromide after various time intervals and estimating the yield of $Ph_3Ge(CH_2)_{17}CH_3$. The initial yield of 79% decreases to 68% after 24 hr, and to 48% after 116 hr at room temperature. After 8 days at room temperature only gummy material is present, although the Gilman colour Test 1 remains positive (H. Gilman and M. B. Hughes, private communication).

Estimation of triphenylgermyl-lithium in ethereal solvents presents the same difficulties as are encountered with organolithium reagents, since both can produce titratable base on hydrolysis, derived from traces of lithium metal in suspension, or solvent cleavage products. An accurate method uses a double-titration technique, in which an aliquot is treated with allyl bromide under nitrogen, hydrolysed after shaking for several minutes, and then titrated with standard acid. This gives the titratable base not due to the germyl-lithium—

$$R_3GeLi + C_3H_5Br \rightarrow R_3GeC_3H_5 + LiBr$$

This value, deducted from the base produced by direct hydrolysis of an aliquot gives the concentration of the germyl-lithium solution (Gilman *et al.*, 1963). Triphenylgermyl-lithium undergoes a hydrogen–metal exchange reaction with "acidic" hydrogen, as in the 9-position of fluorene, and comparative studies based on carbonation to fluorene-9-carboxylic acid show that Ph_3Si^- and Ph_3Ge^- metallate fluorene extensively over 18 hr at 20°C (Ph_3Si, $Ph_3Ge > Ph_3Sn > Ph_3Pb$) (Gilman and Gerow, 1958; Gilman *et al.*, 1962a)—

Tetrabenzylgermane is of some interest in its cleavage by lithium in 1,2-dimethoxyethane. The reaction is inhibited by peroxydic impurities in the solvent, but under carefully controlled conditions it has given tri-benzylgermyl-lithium in about 80% yield—

$$(PhCH_2)_4Ge + 2Li \xrightarrow[\text{C}_4\text{H}_{10}\text{O}_2]{0°-20°C} (PhCH_2)_3GeLi + PhCH_2Li$$

Hydrolysis gives tribenzylgermane, together with some dibenzylgermane and tribenzylmethylgermane. Toluene, ethylbenzene and methylvinyl ether are also produced. There is little doubt that dibenzylgermane is produced via a dilithio-derivative, which is formed by cleavage of a second benzyl—germanium bond by benzyl-lithium—

$$(PhCH_2)_3GeLi + PhCH_2Li \longrightarrow (PhCH_2)_2 + (PhCH_2)_2GeLi_2$$

$$\text{H}_2\text{O}$$

$$(PhCH_2)_2GeH_2$$

The remaining products are all derived from solvent cleavage, either by benzyl-lithium or tribenzylgermyl-lithium—

$$(PhCH_2)_3GeLi + (MeOCH_2)_2 \longrightarrow (PhCH_2)_3GeMe + LiOCH_2CH_2OMe$$

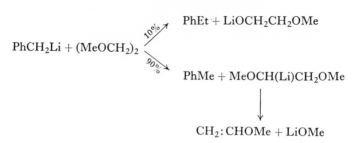

$$CH_2 : CHOMe + LiOMe$$

If tetrabenzylgermane in 1,2-dimethoxyethane is heated with an excess of lithium, a succession of cleavage reactions occurs, involving intermediate lithiobenzylmethylgermanes—

$$(PhCH_2)_3GeLi \xrightarrow{C_4H_{10}O_2} (PhCH_2)_3GeMe \xrightarrow{Li} (PhCH_2)_2Ge(Li)Me \xrightarrow{C_4H_{10}O_2}$$
$$(PhCH_2)_2GeMe_2 \xrightarrow{Li} PhCH_2(Me)_2GeLi \xrightarrow{C_4H_{10}O_2} PhCH_2GeMe_3$$

Hexabenzyldigermane reacts slowly and incompletely with lithium in monoglyme; the main product, tribenzylgermyl-lithium results from cleavage of the germanium—germanium bond, but some cleavage of benzyl–germanium bonds also occurs, since toluene is formed after hydrolysis. The branched-chain compound $(Ph_3Ge)_3GeH$ is of interest, since, with butyl-lithium, metallation and germanium—germanium bond cleavage occur—

$$(Ph_3Ge)_3GeH \xrightarrow{Bu^nLi} (Ph_3Ge)_3GeLi + (Ph_3Ge)_2Bu^nGeLi$$

$$\downarrow MeI$$

$$(Ph_3Ge)_3GeMe + (Ph_3Ge)_2Bu^nGeMe$$

The existence of dialkali metal derivatives of the type R_2GeM_2 is not confined to germanium, and tin analogues, such as Me_2SnNa_2 and Ph_2SnLi_2, have been identified (Kettle, 1959; Schumann *et al.*, 1964a). Sodium naphthalene in monoglyme, which is unreactive towards the germanium—germanium bond in hexaethyldigermane, cleaves cyclic phenylpolyger-manes, $(Ph_2Ge)_{4,5,6}$ (Neumann and Kuehlein, 1967a)—

$$(Ph_2Ge)_n \xrightarrow{NaC_{10}H_8} Ph_2GeNa_2 + Ph_2(Na)GeGe(Na)Ph_2$$

$$\downarrow MeI$$

$$Ph_2GeMe_2 + Ph_2MeGeGeMePh_2$$

A similar reaction giving Ph_2GeNa_2 occurs with sodium in liquid ammonia (Kraus and Brown, 1930), or in low yield by metallation of diphenyl-germane—

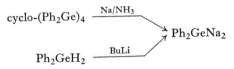

cyclo-$(Ph_2Ge)_4$ $\xrightarrow{Na/NH_3}$

 Ph_2GeNa_2

Ph_2GeH_2 $\xrightarrow{BuLi}$

Potassium in hexamethylphosphotriamide reacts with di-n-butyldichloro-germane in the same way (Bulten and Noltes, 1967)—

$$Bu^n_2GeCl_2 + 4K \xrightarrow{(Me_2N)_3PO} Bu^n_2GeK_2 \xrightarrow{H_2O} Bu^n_2GeH_2 \quad (20\%)$$

The triphenylmethyl group is even more readily cleaved than phenyl by sodium–potassium alloy in ether, and tritylpotassium is not decomposed by the solvent (Brook and Gilman, 1954)—

$$Ph_3GeCPh_3 \xrightarrow[Et_2O]{Na/K} Ph_3GeK + Ph_3CK$$

$$\downarrow CO_2$$

$$Ph_3GeCO_2H + Ph_3CCO_2H$$

The preceding discussion implies that benzyl— and aryl—germanium bonds are more susceptible to nucleophilic attack by alkali metals than are alkyl—germanium bonds. This behaviour is further shown by mixed alkyl-arylgermanes such as triphenyloctadecyl- and triphenyl-2-phenylethyl-germane, both of which react with lithium by cleavage of one phenyl group (Gilman *et al.*, 1959)—

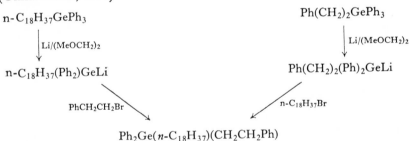

n-$C_{18}H_{37}GePh_3$

$\downarrow$ Li/(MeOCH$_2$)$_2$

n-$C_{18}H_{37}(Ph_2)GeLi$

$Ph(CH_2)_2GePh_3$

$\downarrow$ Li/(MeOCH$_2$)$_2$

$Ph(CH_2)_2(Ph)_2GeLi$

PhCH$_2$CH$_2$Br

n-C$_{18}$H$_{37}$Br

$Ph_2Ge(n$-$C_{18}H_{37})(CH_2CH_2Ph)$

The low reactivity of the germanium—germanium bond in hexaethyl-digermane is reflected in the observation that the mixed germanium–silicon compound is not formed when hexaethyldigermane and triethylbromosilane are refluxed with sodium in tetrahydrofuran (Shackelford *et al.*, 1963).

IV. GROUP II METAL COMPOUNDS

A. Group IIA

There have been no reports of the formation of germanium–beryllium or germanium–calcium compounds though there is no reason to doubt that compounds, such as $R_3GeBeCl$ and R_3GeBeR' might exist, probably as ether or amine complexes. Their properties would be interesting in relation to germyl Grignard reagents, R_3GeMgX.

Strontium and barium in liquid ammonia solution cleave the germanium —germanium bond in hexaphenyldigermane, giving, finally, tetrahydro-furan complexes that decompose at about 40°C (Amberger et $al.$, 1966)—

$$Ph_6Ge_2 \xrightarrow{\text{Sr, NH}_3} (Ph_3Ge)_2Sr \xrightarrow{\text{THF}} (Ph_3Ge)_2Sr(THF)$$

Germyl Grignard reagents

There can be little doubt about the existence of organogermyl Grignard reagents in various ethereal solvents, although none has been isolated or obtained free from other compounds. The first clear evidence was presented by Gilman and Zuech (1961) who studied the reaction between allylmagnesium chloride and triphenylgermane in refluxing tetrahydro-furan—

$$Ph_3GeH + CH_2:CHCH_2MgCl \xrightarrow{\text{THF}} Ph_3GeMgCl + C_3H_6$$

Carbonation of the reaction mixture gave the germyl carboxylic acid in 10% yield, whereas hydrolysis gave a 50% yield of the tetrahydrofuran cleavage product—

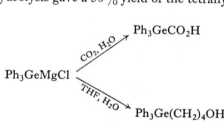

Cyclic ethers, such as tetrahydrofuran, are known to be readily cleaved by reactive organometallic compounds.

In the alkyl- or aryl-ation of germanium tetrachloride by alkyl- and aryl-magnesium halides coupling reactions leading to the digermane R_3GeGeR_3 occur especially in the presence of free magnesium (Glockling and Hooton, 1962b; Quane and Hunt, 1968)—

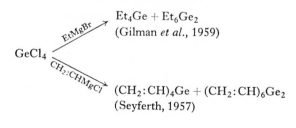

Various reaction paths leading to digermanes have been considered, each of which involves an R_3GeMgX intermediate. Direct reaction of R_3GeX with reactive finely divided free magnesium present in the mixture has been postulated—

$$R_3GeX + Mg \rightarrow R_3GeMgX$$
$$R_3GeX + R_3GeMgX \rightarrow R_6Ge_2$$

The evidence for the formation of germyl Grignard reagents in this way is reasonably well established. For example, carefully filtered solutions of phenylmagnesium bromide and germanium tetrachloride give high yields of tetraphenylgermane, whereas if excess magnesium is present, some 70% of hexaphenyldigermane is produced (Glockling and Hooton, 1966). This is equally true of p- and m-tolyl Grignard reagents, where, with excess magnesium present throughout the reaction, the digermane is formed in quite high yields. Moreover, carbonation or hydrolysis leads to the isolation of the carboxylic acid or hydride—

$$(p\text{-MeC}_6\text{H}_4)_3\text{GeMgBr} \begin{array}{c} \xrightarrow{\text{H}_2\text{O}} (p\text{-MeC}_6\text{H}_4)_3\text{GeH} \\ \xrightarrow{\text{CO}_2} (p\text{-MeC}_6\text{H}_4)_3\text{GeCO}_2\text{H} \end{array}$$

In these two cases it is evident that at least part of the germyl Grignard reagent survives in solution up to a hydrolysis stage. The effect of "reactive" magnesium is quite strikingly illustrated by the reaction of triphenylchlorosilane with cyclohexylmagnesium bromide. In the total absence of free magnesium, only low yields (8%) of disilane, Ph_6Si_2, are produced, whereas with excess magnesium present conversion to the disilane is rapid—

$$Ph_3SiCl + C_6H_{11}MgBr + Mg \longrightarrow Ph_3SiMgBr \xrightarrow{Ph_3SiCl} Ph_6Si_2 \quad (75\%)$$

Reaction of R_3GeX with the Grignard reagent may produce a low equilibrium concentration of R_3GeMgX. The evidence for this route to germyl Grignard reagents is less direct and is based on the observation that the reaction between germanium tetrachloride and sterically hindered Grignard reagents, such as o-tolylmagnesium bromide gives $(o\text{-tolyl})_6Ge_2$, even in the total absence of free magnesium, together with $(o\text{-tolyl})_3GeBr$. By contrast the same reaction is much more rapid with free magnesium present, and gives, after hydrolysis, $(o\text{-tolyl})_6Ge_2$ and the hydride, $(o\text{-tolyl})_3GeH$. This has led to the view that, for sterically hindered R groups, such as o-tolyl, an equilibrium concentration of germyl Grignard reagent is formed—

$$(o\text{-tolyl})_3\text{GeBr} + o\text{-tolylMgBr} \rightleftharpoons (o\text{-tolyl})_3\text{GeMgBr} + o\text{-tolylBr}$$

Isopropylmagnesium chloride is similar, in the sense that, after hydrolysis, tri-isopropylgermane is formed in quite high yield, even in the absence of free magnesium; hence the germyl Grignard reagent is stable in ether solution (Carrick and Glockling, 1966). The formation of germyl Grignard reagents in this way is only expected for organic groups having a large steric requirement, where substitution of the fourth halogen is either slow or does not occur at all. Cyclohexylmagnesium chloride and germanium tetrachloride provide a further example of a relatively stable germyl Grignard reagent, since the hydride $(C_6H_{11})_3GeH$ is formed on hydrolysis and the yield increases with the molar ratio of Grignard reagent. In this case, it has been established that the hydride is not present before hydrolysis (Mendelsohn et al., 1968). A further feature of interest is the indirect evidence that halogen–Grignard exchange reactions can occur at earlier stages of alkylation—

$$Pr^i_2GeCl_2 + Pr^iMgCl \rightarrow ClPr^i_2GeMgCl + Pr^iCl$$

$$ClPr^i_2GeMgCl \xrightarrow{Pr^i_2GeCl_2} ClPr^i_2GeGePr^i_2Cl \xrightarrow{LiAlH_4} (HPr^i_2Ge)_2$$

In the reaction between silicon tetrachloride and lithium, Gilman and Smith (1967) invoke the intermediate $LiSiCl_3$ to account for the ultimate formation of $(Me_3Si)_4Si$ in the reaction—

$$SiCl_4 + Me_3SiCl + Li \rightarrow (Me_3Si)_4Si$$

Reduction of $GeCl_4$ to $GeCl_2$ followed by successive alkylation is a further reasonable route to germyl Grignard reagents—

$$GeCl_4 + 2RMgCl \rightarrow GeCl_2 + 2MgCl_2 + R_2 \text{ (or alkane + alkene)}$$

$$GeCl_2 + 3RMgCl \rightarrow R_3GeMgCl + 2MgCl_2$$

There is no direct evidence in support of this scheme, but it is known that germanium(II) iodide and excess phenyl-lithium give low yields of triphenylgermane (Glockling and Hooton, 1963)—

$$GeI_2 + 3PhLi \longrightarrow Ph_3GeLi \xrightarrow{H_2O} Ph_3GeH$$

Mendelsohn et al. (1968) have studied the alkylation of germanium tetrachloride by sterically hindered Grignard reagents at low temperatures, when high yields of the hydrides R_3GeH are produced after hydrolysis. A modified mechanism has been proposed, in which germanium dichloride is considered to insert into the Grignard reagent, forming $RGe(Cl)_2MgX$, which is then di-alkylated.

Perhaps the most direct evidence for germyl Grignard reagents is the cleavage of a germanium–gold complex by magnesium bromide (p. 199)—

$$Ph_3GeAuPPh_3 + MgBr_2 \rightleftharpoons Ph_3GeMgBr + Ph_3PAuBr$$

B. Group IIB

Zinc, cadmium and mercury form relatively stable bonds with germanium. Zinc and cadmium compounds have been superficially examined and an interesting cleavage of the germanium—cadmium bond by carbon tetrachloride has been reported (Vyazankin *et al.*, 1966c; Amberger *et al.*, 1966)—

$$Ph_3GeNa \xrightarrow[NH_3]{ZnCl_2} (Ph_3Ge)_2Zn \xrightarrow{THF} (Ph_3Ge)_2Zn(THF)_2$$

$$Et_3GeH \xrightarrow{Et_2Cd} (Et_3Ge)_2Cd \xrightarrow{CCl_4} Et_3GeCl + (Et_3Ge)_2CCl_2 + CdCl_2$$

The reaction between triethylgermane and diethylzinc is more complex. At 150°C, ethane, zinc and tetraethylgermane are formed together with a yellow involatile residue, for which a polymeric structure of the type $Et_3GeZn(GeEt_2)_nZnGeEt_3$ has been proposed. Bistriethylgermylcadmium is an air- and water-sensitive compound that has been cleaved by benzyl chloride—

$$(Et_3Ge)_2Cd \xrightarrow{PhCH_2Cl} 2Et_3GeCH_2Ph + CdCl_2$$

It is completely decomposed over 7 hr at 125°C.

Germanium–mercury compounds have received far more study, and are proving of some synthetic importance. Two preparative methods, both of which are probably quite general, have been used (Wiberg *et al.*, 1963; Vyazankin *et al.*, 1963; Eaborn *et al.*, 1967)—

$$\begin{aligned} R_3GeH + Et_2Hg \searrow \\ &\quad\quad (R_3Ge)_2Hg \\ R_3GeX + Hg/Na \nearrow \end{aligned}$$

$$Et_3SiHgEt + Et_3GeH \longrightarrow Et_3SiHgGeEt_3$$

The sodium amalgam reaction is conveniently carried out in cyclohexane at room temperature with total exclusion of air and water; after a period varying from 1 to 10 days, the solution turns yellow and a further week's shaking produces the bistrialkylgermylmercury in some 35% yield.

The lower alkyl complexes sublime readily *in vacuo*, when they give beautiful highly refracting yellow crystals. The methyl complex, $(Me_3Ge)_2Hg$, melts without decomposition at 120°–122°C. Bistriethylgermylmercury is only 10% decomposed after 19 hr at 160°C. In most of its reactions mercury is produced, as in the following illustrations—

$$\begin{aligned} &\qquad\qquad\qquad \xrightarrow{O_2} (R_3Ge)_2O + Hg \\ R_3P \diagdown \quad \diagup GeR_3 \quad \xleftarrow{(R_3P)_2PtCl_2} (R_3Ge)_2Hg \xrightarrow{h\nu} (R_3Ge)_2 + Hg \\ Cl \diagup Pt \diagdown PR_3 \qquad\qquad\qquad \xrightarrow{Li/THF} R_3GeLi \end{aligned}$$

Insertion of dichlorocarbene into the germanium—mercury bond is a reaction of some interest. If dichlorocarbene is generated from the mercury compound $PhHgCCl_2Br$, the overall course of the reaction is given by—

$$(Me_3Ge)_2Hg + 2PhHgCCl_2Br$$

$$\Big\downarrow C_6H_6,\ 70°C$$

$$Me_3GeCl + Me_3GeCCl:CCl_2 + 2PhHgBr + Hg$$

These products may be accounted for if the reaction involves successive insertion of CCl_2 into one of the germanium—mercury bonds (Seyferth *et al.*, 1967b)—

$$(Me_3Ge)_2Hg + 2PhHgCCl_2Br—Me_3GeHgCCl_2CCl_2GeMe_3 + 2PhHgBr$$

$$\nearrow$$

$$Me_3GeCCl:CCl_2 + Me_3GeHgCl$$

$$\Big\downarrow$$

$$Me_3GeCl + Hg$$

Ethylene dibromide cleaves germanium—mercury bonds with formation of mercury, and this also is likely to proceed via an unstable mercury–halide intermediate—

$$(Et_3Ge)_2Hg + C_2H_4Br_2 \longrightarrow Et_3GeBr + Et_3GeHgBr + C_2H_4$$

$$\nearrow$$

$$Et_3GeBr + Hg$$

Acetic acid behaves similarly, again giving mercury—

$$(Et_3Ge)_2Hg + HOAc \rightarrow Et_3GeH + Et_3GeOAc + Hg$$

A wide range of unsaturated compounds react rapidly with bistriethyl-germylmercury (Kuehlein *et al.*, 1967)—

$$RO_2CC:CCO_2R + (Et_3Ge)_2Hg \rightarrow Hg + Et_3GeCCO_2R$$
$$\|$$
$$Et_3GeCCO_2R$$

It is unlikely that initiation of these reactions involves free radicals, although radicals must feature in the propagation stages, since e.s.r. signals are observed.

V. GROUP III METAL COMPOUNDS

Only one germanium–boron compound has been fully characterized in the form of a complex anion with tetraco-ordinate boron—

$$Ph_3GeLi + Ph_3B \longrightarrow Li[Ph_3GeBPh_3]$$

$$\downarrow Me_4NCl$$

$$Me_4N[Ph_3GeBPh_3]$$

The complex is rapidly hydrolysed in aqueous solution, but evidently resists solvolysis by methanol. An earlier experiment on the reaction between triphenylgermyl-lithium and dibutylboron chloride gave some evidence for the formation of a germanium–boron compound, possibly $(Ph_3Ge)_3B \cdot OEt_2$ (Seyferth et al., 1961; Booth and Kraus, 1952).

The only evidence for the existence of germanium–aluminium compounds is indirect, and is derived from the isolation of digermanes in the reaction of aluminium alkyls with germanium tetrachloride (p. 149). No gallium or indium complexes with germanium have been reported, but one thallium complex has been described (Kruglaya et al., 1967). Triethylgermane and triethylthallium react at 100°C to give ethane, a little thallium and tristriethylgermylthallium as a red oil—

$$3Et_3GeH + Et_3Tl \xrightarrow{100°C} (Et_3Ge)_3Tl + 3C_2H_6$$

Its thermal stability is considerable, and at 170°C it decomposes to thallium and hexaethyldigermane. Hydrolysis is reported as giving thallium, triethylgermane and hexaethyldigermoxane. Other reactions that illustrate the high reactivity of this compound are as follows—

$$(Et_3Ge)_3Tl \begin{cases} \xrightarrow{Hg, 20°C} (Et_3Ge)_2Hg \\ \xrightarrow{C_2H_4Br_2} C_2H_4 + TlBr + Et_3GeBr \\ \xrightarrow{Et_3SnH} Et_3GeH + Et_6Sn_2 + Tl \end{cases}$$

VI. GROUP V METAL COMPOUNDS

Germanium–arsenic compounds have not been reported, but tristrimethylgermylantimony is obtained as a low-melting solid by the reaction (Amberger and Salazar, 1967)—

$$Sb + 3Li \xrightarrow{NH_3} Li_3Sb \xrightarrow{Me_3GeCl} (Me_3Ge)_3Sb$$

One germanium—antimony bond may be cleaved by benzyl bromide—

$$(Et_3Ge)_3Sb + PhCH_2Br \rightarrow Et_3GeBr + (Et_3Ge)_2SbCH_2Ph$$

Tristriethylgermylbismuth is obtained by an exchange reaction with the analogous silyl compound, and this in turn may be converted into the tin analogue—

$$(Et_3Si)_3Bi + Et_3GeH \xrightarrow{180°C} 3Et_3SiH + (Et_3Ge)_3Bi$$

$$\searrow Et_3SnH$$

$$(Et_3Sn)_3Bi$$

The general reaction between a germanium hydride and metal alkyl has also been applied to the formation of a germanium—bismuth bond; in this case a mixture of products may be obtained (Kruglaya *et al.*, 1966)—

$$Et_3GeH + Et_3Bi \rightarrow Et_3GeBiEt_2 + (Et_3Ge)_2BiEt + C_2H_6$$

Like the group IIA and IIB compounds, these are light sensitive and react rapidly with oxygen and water.

VII. TRANSITION-METAL COMPLEXES

Most transition metals form σ-bonds to one or other of the group IV elements (silicon, germanium, tin and lead). The information available on some of these types is extremely limited, but it does appear that, with few exceptions, the metal—metal bonds are of quite high thermodynamic stability, though their chemical reactivity varies enormously. High chemical reactivity is only observed for those transition metals that readily increase their co-ordination number, usually by "oxidative" reactions, thereby giving intermediates of low stability. In this way reactions occur that require a surprisingly low activation energy. For example, the metal—metal bond in the complex $[\pi\text{-}C_5H_5(CO)_2Fe]_2SnPh_2$ is so unreactive that hydrogen chloride in carbon tetrachloride solution simply cleaves the tin—phenyl bonds forming $[\pi\text{-}C_5H_5(CO)_2Fe]_2SnCl_2$. By contrast, one platinum—germanium bond in $(Et_3P)_2Pt(GePh_3)_2$ is rapidly cleaved by molecular hydrogen at 1 atm and 0°C. In other complexes, cleavage reactions by polar reagents follow what appears to be the polarity of the metal—metal bond (e.g., $\overset{\delta-}{Mo}$—$\overset{\delta+}{Ge}$, $\overset{\delta+}{Au}$—$\overset{\delta-}{Ge}$).

As with σ-bonded transition metal–carbon compounds, it seems that the heavier transition metals form the more stable complexes (Au $\gg$ Ag > Cu, Pt $\gg$ Pd). These differences may be attributed to various inter-related factors, including electronegativity differences, the effectiveness of orbital overlap and the extent of d_π–d_π character to the metal—metal bond (Coffey *et al.*, 1964). In almost all of the stable compounds so far reported, the transition metal is in a low oxidation state. The effect of co-ordination number is apparent with copper and silver complexes, which are most stable when the transition metal is tetraco-ordinate, as in $(Ph_3P)_3CuGePh_3$. A further factor that probably affects both stability and reactivity is the charge on the complex. Again, little information is available, but several anionic and cationic complexes of various transition metals with germanium show marked differences from similar neutral complexes.

The effect of π-bonding ligands in stabilizing metal—carbon σ-bonds has been extensively studied, and the thermal stability of transition metal—group IV metal bonds appears even more sensitive to changes in π-bonding ligands attached to the transition metal. The remarkable stability and range of complexes formed by the $[SnCl_3]^-$ ion in combination with transition metals, e.g., $[Pt(SnCl_3)_5]^{4-}$, has been considered in relation to its σ-donor and π-acceptor characteristics (Young et al., 1964; Cramer et al., 1965; Lindsey et al., 1966; Parshall, 1966). For the series silicon, germanium, tin and lead, there is some indication that maximum stability occurs with germanium and tin, but there could be confusion between stability and reactivity. Thus $Ph_3PAuSiPh_3$ and $(Et_3P)_2Pt(H)SiPh_3$ are rather air and light sensitive, whereas the germanium analogues are not. With organo-tin and -lead compounds, some reactions occur by cleavage of the tin—carbon or lead—carbon rather than the metal—metal bond.

A. Preparative methods

A wide range of essentially standard experimental methods are available for forming transition metal—group IV metal bonds; the best method for a given compound being most dependent on the particular transition metal. These may be classified under five headings.

Hydrogen and hydrogen halide elimination reactions

$$2(CO)_5MnH + GeH_4 \rightarrow [(CO)_5Mn]_2GeH_2 + 2H_2$$
$$(CO)_5MnCl + HGeCl_3 \rightarrow (CO)_5MnGeCl_3 + HCl$$

Elimination of alkali metal halides

$$\pi\text{-}C_5H_5(CO)_3MoNa + R_3GeCl \rightarrow \pi\text{-}C_5H_5(CO)_3MoGeR_3 + NaCl$$
$$Ph_3GeLi + Ph_3PAuCl \rightarrow Ph_3PAuGePh_3 + LiCl$$

From metal halides and metal–metal complexes

$$(R_3P)_2PtCl_2 + (Me_3Ge)_2Hg \rightarrow (R_3P)_2Pt(Cl)GeMe_3 + Me_3GeCl + Hg$$
$$(CO)_4CoSiMe_3 + GeF_4 \rightarrow (CO)_4CoGeF_3 + Me_3SiF$$

"Insertion" reactions

$$Co_2(CO)_8 + GeI_2 \rightarrow (CO)_4CoGe(I)_2Co(CO)_4$$

Cleavage of organogermanium amides by metal hydrides

$$R_3GeNR_2' + (R_3P)_2Pt(H)Cl \rightarrow (R_3P)_2Pt(Cl)GeR_3 + R_2'NH$$

B. Titanium and zirconium

Germanium forms stable compounds with titanium in its +3 and +4 oxidation states, and zirconium has yielded the zirconium(IV) complex, $(\pi\text{-}C_5H_5)_2(Cl)ZrGePh_3$ (Coutts and Wailes, 1968)—

$$(\pi\text{-}C_5H_5)_2TiCl_2 + Ph_3GeLi \xrightarrow{\text{THF}} (\pi\text{-}C_5H_5)_2(Cl)TiGePh_3$$

$$(\pi\text{-}C_5H_5)_2TiCl + Ph_3GeLi \xrightarrow{\text{THF}} (\pi\text{-}C_5H_5)_2(THF)TiGePh_3$$

The titanium(IV) complex is unusual in being bright green although it is diamagnetic. Titanium(III) complexes are also green, and have only been obtained as solvates; their paramagnetism corresponds to one unpaired electron with no evidence of spin–spin interaction. They react with chloroform to give titanium(IV) chloro-complexes.

C. Chromium, molybdenum and tungsten

The chromium complex $\pi\text{-}C_5H_5(CO)_3CrGePh_3$ has been obtained as a yellow–green solid that decomposes in an inert atmosphere over several weeks at room temperature (Patil and Graham, 1966). Both chromium and tungsten complexes of germanium result from insertion reactions between the dianions, $[M_2(CO)_{10}]^{2-}$ and germanium(II) iodide. The products of these reactions are rapidly oxidized in solution, but as crystalline solids they can be handled in air (Ruff, 1967b)—

$$[(Ph_3P)_2N]_2Cr_2(CO)_{10} + GeI_2 \rightarrow [(Ph_3P)_2N]_2[(CO)_5CrGe(I)_2Cr(CO)_5]$$

Trialkylgermyl complexes of molybdenum and tungsten have been more thoroughly examined. They are formed in high yield under mild conditions by the reaction—

$$\pi\text{-}C_5H_5(CO)_3MNa + R_3GeBr \rightarrow \pi\text{-}C_5H_5(CO)_3MGeR_3$$

Their thermal stability is considerable, and they sublime readily *in vacuo* (when R = Me or Et) as colourless solids. In reactions with polar reagents they show the bond polarity $\overset{\delta-}{M}\text{—}\overset{\delta+}{Ge}$.

Prolonged heating of the molybdenum or tungsten complexes $\pi\text{-}C_5H_5(CO)_3MoGeMe_3$ and $\pi\text{-}C_5H_5(CO)_3WGeEt_3$ at 200°C *in vacuo* produces less than 1 mole of carbon monoxide together with the carbonyl dimers $[\pi\text{-}C_5H_5(CO)_3M]_2$ and some evidence for trace quantities of $R_3GeC_5H_5$.

Molybdenum complexes are more readily oxidized by air than tungsten analogues, but both are stable to water. Although as crystalline solids the molybdenum complexes rapidly acquire a red–brown surface coating in oxygen produced by superficial oxidation, in solution oxidation is much more rapid, giving after 18 hr a deep-blue insoluble molybdenum-containing material that is slowly oxidized to MoO_3. The limited analytical data on this intermediate suggest a composition π-$C_5H_5(CO)_3(O)_4MoGeEt_3$. It is of some interest that carbon dioxide is formed in the overall oxidation process, but the fate of the cyclopentadienyl group has not been determined—

$$\pi\text{-}C_5H_5(CO)_3MoGeR_3 \xrightarrow[C_6H_6]{O_2} CO + CO_2 + (R_3Ge)_2O + MoO_3$$

Solutions of tungsten complexes oxidize in the same way, but some 80% may be recovered unchanged after remaining in contact with excess oxygen for 40 days.

Some ligand-exchange reactions have been reported that do not cleave the metal—metal bond. Diethylphosphine will displace one carbonyl group, whereas pyridine shows little evidence of reaction—

$$\pi\text{-}C_5H_5(CO)_3MoGeR_3 + Et_2PH \rightarrow \pi\text{-}C_5H_5(CO)_2(Et_2PH)MoGeR_3 + CO$$

The mixed phosphine-carbonyl is apparently much more air sensitive than the tricarbonyl complex. Triphenylphosphine behaves quite anomalously giving evidence for an intermediate (11), which reversibly loses carbon monoxide—

$$\pi\text{-}C_5H_5(CO)_3MoGeR_3 + Ph_3P$$
$$\updownarrow$$
$$\pi\text{-}C_5H_5(CO)_3(Ph_3P)MoGeR_3$$
$$(11)$$
$$\updownarrow$$
$$\pi\text{-}C_5H_5(CO)_2(Ph_3P)MoGeR_3 + CO$$

The most interesting aspect of this reaction is the structure of the proposed intermediate (11). It could be an octaco-ordinate complex, but the magnetic resonance changes of the C_5H_5 protons as the reaction proceeds suggest that the bonding of the C_5H_5 group is altered (possibly becoming a π-allyl ligand) in the intermediate. Similar reactions on organomolybdenum complexes proceed to an acyl derivative, but the reaction of triphenylphosphine with the germyl molybdenum complex gives no evidence for the grouping $MoCOGeR_3$.

Some cleavage reactions have been reported for germanium–molybdenum and germanium–tungsten complexes. Hydrogen chloride, ethylene dibromide and ethyl bromide all cleave the metal—metal bond, though the last of these is clearly a rather complicated free-radical reaction.

$$\pi\text{-}C_5H_5(CO)_3WGeMe_3 \begin{cases} \xrightarrow{HCl} & \pi\text{-}C_5H_5(CO)_3WH + Me_3GeCl \\ \xrightarrow{C_2H_4Br_2} & \pi\text{-}C_5H_5(CO)_3WBr + Me_3GeBr + C_2H_4 \\ \xrightarrow{EtBr} & [\pi\text{-}C_5H_5(CO)_3W]_2 + Me_3GeBr + C_2H_4 + C_2H_6 + C_4H_{10} \end{cases}$$

Excess of iodine in benzene solution reacts in a rather unusual way by cleaving the metal—metal bond with displacement of one carbonyl group—

$$\pi\text{-}C_5H_5(CO)_3WGeR_3 + 2I_2 \rightarrow \pi\text{-}C_5H_5(CO)_2(I_2)WI + R_3GeI + CO$$

Mercuric chloride and magnesium bromide also cleave the metal—metal bond (Carrick and Glockling, 1968)—

$$\pi\text{-}C_5H_5(CO)_3MGeR_3 \begin{cases} \xrightarrow{HgCl_2} & \pi\text{-}C_5H_5(CO)_3MHgCl + R_3GeCl \\ \xrightarrow{MgBr_2} & \pi\text{-}C_5H_5(CO)_3MMgBr(THF) + R_3GeBr \end{cases}$$

Trichlorogermyl complexes of chromium, molybdenum and tungsten have been obtained as complex anions by the reaction (Ruff, 1967a)—

$$Ph_4As^+[GeCl_3]^- + M(CO)_6 \begin{cases} \longrightarrow & Ph_4As[(CO)_5MGeCl_3] + CO \\ \longrightarrow & [Ph_4As]_2[CO)_4M(GeCl_3)_2] + 2CO \end{cases}$$

When heated with triphenylphosphine the germanium–molybdenum complex decomposes in two stages—

$$Ph_4As[(CO)_5MoGeCl_3] \xrightarrow{Ph_3P} Ph_4As[(CO)_4(Ph_3P)MoGeCl_3] + CO$$

$$\downarrow Ph_3P$$

$$[Ph_4As][GeCl_3] + cis\text{-}(CO)_4Mo(PPh_3)_2$$

Although molybdenum- and tungsten-hexacarbonyls are unreactive towards germanium tetrachloride, the bipyridyl carbonyls form heptacoordinate complexes that are non-electrolytes, stable to air and light (Kummer and Graham, 1968)—

$$(bipy)(CO)_4Mo + GeCl_4 \rightarrow (bipy)(CO)_3ClMoGeCl_3 + CO$$

D. Manganese and rhenium

Manganese and rhenium complexes with germanium are most readily prepared by the reaction (Seyferth *et al.*, 1962a; Nesmeyanov *et al.*, 1966a)—

$$(CO)_5MNa + R_3GeBr \rightarrow (CO)_5MGeR_3$$
$$(CO)_5MNa + R_2GeBr_2 \rightarrow [(CO)_5M]_2GeR_2$$

Related preparative methods have yielded germanium-halide and -hydride complexes (Massey *et al.*, 1963; Nesmeyanov *et al.*, 1965a)—

$$(CO)_5MnCl + HGeCl_3 \xrightarrow[60°C]{THF} (CO)_5MnGeCl_3$$

$$(CO)_5MnH + GeH_4 \xrightarrow[15°C]{THF} (CO)_5MnGe(H)_2Mn(CO)_5$$

The second of these reactions, which is almost quantitative, is rather complex mechanistically, since it gives no evidence for $H_3GeMn(CO)_5$ or $HGe[Mn(CO)_5]_3$. Possibly reduction of monogermane to germanium dihydride proceeds metal—metal bond formation.

A further reaction, so far applied to silicon and tin, starts from manganese- or rhenium-carbonyl (Jetz *et al.*, 1966)—

$$Mn_2(CO)_{10} + 2Ph_3SiH \rightarrow 2(CO)_5MnSiPh_3 + H_2$$

All of the manganese–germanium complexes are evidently stable in air. In the triphenylgermyl complex $Ph_3GeMn(CO)_5$, the germanium—manganese bond length is 2·54 Å (Kilbourn *et al.*, 1965). The silyl–manganese compound, $H_3SiMn(CO)_5$ forms complexes with tertiary amines, which have been tentatively formulated as ionic solids (Aylett and Campbell, 1965)—

$$H_3SiCo(CO)_4 + 2py \rightarrow [H_3Si(py)_2]^+[Co(CO)_4]^-$$

Studies on thermal stability are difficult to correlate. For the complexes $Ph_3MMn(CO)_5$ (M = Si, Sn, Pb), Gorsich (1962) found the lead compound less stable than the tin analogue, which only lost carbon monoxide at 195°C. Even $Me_3PbMn(CO)_5$ is sufficiently stable for distillation *in vacuo*. In solution, the manganese–germanium complexes undergo slow aerial oxidation with evolution of carbon monoxide, and the rate is accelerated by light. Also in compounds of the type $R_2M[Mn(CO)_5]_2$, the oxidative stability is lower than in $R_3MMn(CO)_5$ complexes.

Two further types of reaction have been studied, both of which reflect the high stability of the germanium—manganese bond. Neutral ligand replacement reactions on both manganese and rhenium complexes result in displacement of one carbonyl group, e.g.—

$$Ph_3GeRe(CO)_5 + Ph_3P \rightarrow Ph_3GeRe(CO)_4PPh_3 + CO$$

Triphenylarsine and triphenylstibine also react, though less readily. The products are crystalline and air stable. Chlorination of $Ph_3GeMn(CO)_5$ occurs smoothly in carbon tetrachloride, with cleavage of the phenyl groups from germanium. The resulting trichlorogermyl complex is more stable to air and light than the triphenyl analogue, and this seems to be generally true—

$$Ph_3GeMn(CO)_5 + 3Cl_2 \xrightarrow{CCl_4} Cl_3GeMn(CO)_5 + 3PhCl$$

Tin–manganese compounds are chlorinated in the same way. In lead–manganese complexes, the lead—manganese bond is quite readily cleaved by hydrogen chloride at 25°C, whereas the germanium complex $Ph_3GeMn(CO)_5$ is unaffected by hydrogen chloride at 76°C in carbon tetrachloride solution. Several chlorination and bromination reactions have been carried out selectively by varying the experimental conditions, and in this way cleavage of 1,2 or 3 phenyl groups or the metal—metal bond, may be effected. For example, with the rhenium complex $Ph_3GeRe(CO)_5$, all combinations of bromination products, from $Br_3GeRe(CO)_5$ to $Ph_2BrGeRe(CO)_5$, may be obtained in high yield. Iodine at 180°C causes complete decomposition of the germanium hydride complex—

$$[(CO)_5Mn]_2GeH_2 \xrightarrow[180°C]{I_2} 10\ CO + H_2 + GeI_4 + 2MnI_2$$

E. Iron, ruthenium and osmium

Only iron–germanium compounds have so far been reported, but it is highly likely that ruthenium and osmium will behave similarly. Iron–germanium complexes have been made by insertion of germanium di-iodide into cyclopentadienyliron dicarbonyl dimer, or from germanium tetrahalides and the sodium salt of the iron complex. The former reaction takes place in refluxing benzene (Flitcroft et al., 1966), and is, formally, analogous to carbene insertion reactions—

$$[\pi\text{-}C_5H_5(CO)_2Fe]_2 + GeI_2 \searrow$$
$$[\pi\text{-}C_5H_5(CO)_2Fe]_2GeI_2$$
$$2\pi\text{-}C_5H_5(CO)_2FeNa + GeI_4 \nearrow$$

The corresponding dichloride, $[\pi\text{-}C_5H_5(CO)_2Fe]_2GeCl_2$ has been made from trichlorogermane, although the reactive species leading to its formation may be germanium dichloride (Nesmeyanov et al., 1966b)—

$$[\pi\text{-}C_5H_5(CO)_2Fe]_2 + HGeCl_3 \rightarrow$$
$$[\pi\text{-}C_5H_5(CO)_2Fe]_2GeCl_2 + \pi\text{-}C_5H_5(CO)_2FeGeCl_3$$
$$\pi\text{-}C_5H_5(CO)_2FeCl + HGeCl_3 \rightarrow \pi\text{-}C_5H_5(CO)_2FeGeCl_3 + HCl$$

All the compounds reported are air-stable solids. An X-ray examination of $[\pi\text{-}C_5H_5(CO)_2Fe]_2GeCl_2$ shows a distorted tetrahedral distribution of groups about the germanium atom, the iron–germanium–iron angle being 128° and the chlorine–germanium–chlorine angle 96° (Bush and Woodward

1967). The germanium–iron bond length is 2·36 Å and if the covalent radius of germanium is taken as 1·21 Å, this gives the extremely small covalent radius for iron of 1·14 Å. The germanium–chlorine bond length of 2·26 Å is significantly longer than in germanium tetrachloride (2·09 Å) or simple chlorogermanes (2·11–2·15 Å). Furthermore, an examination of the carbonyl stretching frequencies in solution is compatible with molecular symmetry C_1, whereas in the crystal it is C_2. Methylation of this compound produces a dimethyl derivative of greater air sensitivity, and metal hydride reduction has been carried out without affecting the germanium–iron bonds. The dihydride is slowly chlorinated by chloroform—

$$[\pi\text{-}C_5H_5(CO)_2Fe]_2GeCl_2 \xrightarrow{MeLi} [\pi\text{-}C_5H_5(CO)_2Fe]_2GeMe_2$$
$$\xrightarrow{NaBH_4} [\pi\text{-}C_5H_5(CO)_2Fe]_2GeH_2$$

Comparison of the carbonyl stretching frequencies in $[\pi\text{-}C_5H_5(CO)_2\text{-}Fe]_2GeCl_2$ and $[\pi\text{-}C_5H_5(CO)_2Fe]_2GeMe_2$ shows them to be higher in the dichloride. Since chlorine will be more effective than methyl groups in removing negative charge, the tendency of iron to take part in $d\pi$–$p\pi$ bonding with the carbonyl groups will be reduced in the dichloride and hence lead to a higher carbon–oxygen bond order.

A related reaction, not yet applied to germanium, is that of tributyltin chloride and iron pentacarbonyl. This is a complex reaction producing a 4-membered iron–tin ring system together with dibutyl ketone, carbon monoxide and ferrous chloride—

$$Bu_3SnCl + Fe(CO)_5 \longrightarrow (CO)_4Fe \underset{\underset{Bu_2}{Sn}}{\overset{\overset{Bu_2}{Sn}}{\diagup\diagdown}} Fe(CO)_4$$

Ruthenium–tin complexes have been obtained by this type of reaction, and ruthenium–silicon complexes are formed from the silane and ruthenium carbonyl (Cotton *et al.*, 1967a, b)—

$$Me_3SiH + Ru_3(CO)_{12} \rightarrow Me_3SiRu(CO)_4Ru(CO)_4SiMe_3$$

Iron–germanium complexes, such as $Ph_4As[(CO)_4FeGeCl_3]$, analogous to the chromium, molybdenum and tungsten compounds have been described.

F. Cobalt, rhodium and iridium

All three metals form complexes with the group IV elements, and the first of these was reported by Chalk and Harrod (1965). Tri-substituted silanes cleave dicobalt octacarbonyl to give the silicon–cobalt complex and cobalt carbonyl hydride, which then decomposes to hydrogen with regeneration of the octacarbonyl—

$$Co_2(CO)_8 + R_3SiH \rightarrow (CO)_4CoSiR_3 + HCo(CO)_4$$
$$2HCo(CO)_4 \rightarrow H_2 + Co_2(CO)_8$$

Cobalt carbonyl or its anion reacts similarly with germanium di- and tetra-halides—

$$NaCo(CO)_4 + GeCl_4 \rightarrow (CO)_4CoGeCl_3$$
$$Co_2(CO)_8 + GeCl_4 \rightarrow (CO)_4CoGe(Cl)_2Co(CO)_4$$
$$Co_2(CO)_8 + GeI_2 \rightarrow (CO)_4CoGe(I)_2Co(CO)_4$$

The mechanism of the germanium di-iodide insertion reaction is not clear; manganese carbonyl, $Mn_2(CO)_{10}$, which does not have bridging carbonyl groups fails to react with stannous chloride, but the complex $[Ph_3P(CO)_3Co]_2$, which is metal—metal bonded, can be induced to react, though under more vigorous conditions. These reactions take place readily in tetrahydrofuran, and the complexes, which are thermally stable, are oxidized by air (Patmore and Graham, 1966, 1967). Triphenylgermyl– and trifluorogermyl–cobalt complexes have been prepared by the reactions (Baay and MacDiarmid, 1967)—

$$(CO)_4CoNa + Ph_3GeBr \rightarrow (CO)_4CoGePh_3 + NaBr$$
$$(CO)_4CoSiMe_3 + GeF_4 \rightarrow (CO)_4CoGeF_3 + Me_3SiF$$

π-Cyclopentadienyldicarbonylcobalt combines with germanium tetra-halides in the same way as molybdenum complexes, with the formation of one or two germanium—cobalt bonds, depending on the conditions used (Kummer and Graham, 1968)—

$$\pi\text{-}C_5H_5(CO)_2Co \xrightarrow{GeCl_4} \pi\text{-}C_5H_5(CO)(Cl)CoGeCl_3 + \pi\text{-}C_5H_5(CO)Co(GeCl_3)_2$$

Both rhodium(I) and iridium(I) halides add trimethylgermane to yield yellow octahedral complexes (F. Glockling, G. C. Hill and M. Wilbey, unpublished observations)—

$$(Ph_3P)_3RhCl + Me_3GeH \rightarrow (Ph_3P)_3Rh(H)(Cl)GeMe_3 \text{ (unstable)}$$
$$(Ph_3P)_2(CO)IrCl + Me_3GeH \rightarrow (Ph_3P)_2(CO)Ir(H)_2GeMe_3$$

G. Nickel, palladium and platinum

1. Nickel

Only one nickel–group IV metal complex $[\pi\text{-}C_5H_5(CO)Ni]_2SnCl_2$, has been described (Patmore and Graham, 1966). The reaction between triphenylgermyl-lithium and either the planar $(Et_3P)_2NiCl_2$ or the tetrahedral complex $(Ph_3P)_2NiCl_2$ has given only hexaphenyldigermane and tarry materials.

2. *Palladium*

Of the group IV elements, only palladium–germanium complexes have so far been described (Brooks and Glockling, 1966, 1967). As with carbon compounds, such as $(Et_3P)_2PdMe_2$, the thermal stability is much lower than with platinum analogues (Calvin and Coates, 1960).

trans-Bistriethylphosphinepalladium dichloride and triphenylgermyl-lithium react in monoglyme solution to give the PdGe$_2$ complex—

$$trans\text{-}(Et_3P)_2PdCl_2 + 2Ph_3GeLi \rightarrow (Et_3P)_2Pd(GePh_3)_2 + 2LiCl$$

Isolation of the product is difficult owing to its thermal instability in solution; decomposition occurs in toluene solution even at $-20°C$, but the germanium—palladium bonds are stable to water, and to oxygen at $-40°C$. Other palladium dihalide complexes, such as $(Ph_3P)_2PdCl_2$, (bipy)PdCl$_2$, and *cis*-$(Ph_2PCH_2CH_2PPh_2)PdCl_2$, also react with triphenylgermyl-lithium, but without yielding isolable germanium–palladium complexes. This may be due to experimental difficulties and does not necessarily imply an enhanced stability for the triethylphosphine complex.

The stereochemistry of the complex $(Et_3P)_2Pd(GePh_3)_2$ is not known. It is pale yellow, which, by analogy with similar platinum compounds, suggests a *trans* configuration, although in solution it may exist as an equilibrium mixture of *cis* and *trans* forms. Thermal decomposition of the solid is apparent at $97°C$, and at $107°C$ decomposition is complete, giving a variety of products—

$$(Et_3P)_2Pd(GePh_3)_2$$
$$\downarrow 97°\text{--}107°C$$
$$Pd + Ph_6Ge_2 + Ph_4Ge + Et_3P + C_6H_6 + C_2H_4 + H_2$$

This is clearly a complex radical decomposition, possibly initiated by the formation of Ph$_3$Ge radicals, which in addition to dimerizing also abstract a phenyl radical from another Ph$_3$Ge group. Thus there must be other phenylgermanium compounds or germanium metal unaccounted for. The formation of ethylene must be attributed to cleavage of carbon—phosphorus bonds.

Other reactions of this germanium–palladium complex are as follows—

$$\text{(Et}_3\text{P)}_2\text{Pd(GePh}_3)_2 \begin{cases} \xrightarrow[\substack{H_2, \\ 100 \text{ atm}}]{} \text{(Et}_3\text{P)}_2\text{Pd(H)GePh}_3 + \text{Ph}_3\text{GeH} \\ \xrightarrow{C_2H_4Br_2} \text{trans-(Et}_3\text{P)}_2\text{PdBr}_2 + \text{C}_2\text{H}_4 + 2\text{Ph}_3\text{GeBr} \\ \xrightarrow{2KCN} \text{K}_2\,[\text{(CN)}_2\text{Pd(GePh}_3)_2] + 2\text{Et}_3\text{P} \end{cases}$$

These reactions are discussed in greater detail under platinum, but whereas replacement of triethylphosphine by other neutral ligands has been unsuccessful, replacement by cyanide occurs readily giving the complex anion $[\text{(CN)}_2\text{Pd(GePh}_3)_2]^{2-}$, which is evidently of greater thermal stability than the phosphine complex, even when in solution.

The reaction with an excess of hydrogen chloride in ether is of interest in relation to platinum analogues, in that only two products are obtained—

$$\text{(Et}_3\text{P)}_2\text{Pd(GePh}_3)_2 + 2\text{HCl} \rightarrow 2\text{Ph}_3\text{GeH} + \text{(Et}_3\text{P)}_2\text{PdCl}_2$$

However, when the products, triphenylgermane and *trans*-bistriethyl-phosphinepalladium dichloride, are heated together, extensive decomposition occurs with the formation of the palladium hydridochloride and triphenylchlorogermane. A modification of this method using trimethylgermane at 40°C, so that all by products are readily volatile at room temperature, gives the beautifully crystalline palladium hydride—

$$\textit{trans-}\text{(Et}_3\text{P)}_2\text{PdX}_2 + \text{Me}_3\text{GeH} \rightarrow$$
$$\textit{trans-}\text{(Et}_3\text{P)}_2\text{Pd(H)X} + \text{Me}_6\text{Ge}_2 + \text{Me}_3\text{GeCl} + \text{H}_2$$

3. Platinum

Platinum forms complexes with all of the group IV elements (Lindsey *et al.*, 1966; Parshall, 1966; Cross, 1967; Baird, 1967). Silicon and lead complexes, which have been obtained by the reactions—

$$\textit{cis-} \text{ or } \textit{trans-}\text{(Et}_3\text{P)}_2\text{PtCl}_2 + \text{Ph}_3\text{SiLi} \rightarrow \text{(Et}_3\text{P)}_2\text{Pt(H)SiPh}_3$$
$$\textit{trans-}\text{(Ph}_3\text{P)}_2\text{Pt(H)Cl} + \text{Ph}_3\text{PbNO}_3 \rightarrow \text{(Ph}_3\text{P)}_2\text{Pt(Cl)PbPh}_3$$

are less stable than those of germanium and tin. The silicon compound prepared by the above reaction is described as air and light sensitive. Its formation is rather strange, and may result from hydrolysis of a lithio-platinum intermediate produced by halogen–metal exchange—

$$\text{(Et}_3\text{P)}_2\text{Pt(Cl)SiPh}_3 + \text{Ph}_3\text{SiLi} \longrightarrow \text{(Et}_3\text{P)}_2\text{Pt(Li)SiPh}_3 + \text{Ph}_3\text{SiCl}$$
$$\xleftarrow{\text{H}_2\text{O}}$$
$$\text{(Et}_3\text{P)}_2\text{Pt(H)SiPh}_3$$

The triphenyltin complex, $\text{(Ph}_3\text{P)}_2\text{Pt(Cl)SnPh}_3$, is stable to about 278°C, and, when heated in a polar solvent, decomposes by transfer of a phenyl group from tin to platinum with cleavage of the platinum—tin bond. The fate of the residual Ph_2Sn moiety has not been determined with certainty. This type of rearrangement resembles the well known 1,2-shifts of organic chemistry, and migration of organic groups across a metal—metal bond are well established as occurring under electron impact (Chambers and Glockling, 1968). Reference has already been made to the many platinum–$SnCl_3$ complexes, which include both neutral and anionic compounds (Lindsey *et al.*, 1966; Parshall, 1966).

Platinum–germanium complexes have been obtained by two main reactions (Cross and Glockling, 1965b, 1965c; Glockling and Hooton, 1967, 1968; Brooks *et al.*, 1968).

cis- or trans-$(R_3P)_2PtX_2 + 2Ph_3GeLi \rightarrow$ cis- or trans-$(Et_3P)_2Pt(GePh_3)_2$

This preparative method is clearly a two-stage process, but it appears that the rate of triphenylgermylation of the intermediate, $(R_3P)_2Pt(Cl)GePh_3$, is greater than the first stage, so that if equimolar ratios of reactants are used, the products are unreacted material and the bistriphenylgermylplatinum complex. This is not unexpected, since, if the intermediate is the *trans* isomer the chloride will be extremely labile due to the high *trans* effect of the Ph_3Ge group. *Cis* and *trans* forms of the bistriphenylgermyl complexes have been obtained, the *cis* form being colourless and the *trans* form yellow. There is evidence that in solution an equilibrium is established between the isomeric forms, since the colourless *cis* isomers give bright-yellow solutions in benzene or acetone from which the *cis* or *trans* form may be crystallized. Both forms are stable to air, water, ethanol and ethanolic potassium hydroxide. The *cis* triethylphosphine complex $(Et_3P)_2Pt(GePh_3)_2$ gives an apparent dipole moment of only 4·45D, and the *trans* isomer a moment of 2·4D, whereas the chelating phosphine complex (**12**), which necessarily has a *cis* configuration, has a dipole moment of 8·85D in benzene solution—

(**12**)

Triphenylgermyl-lithium and the platinum di-iodide complex react in a more complicated manner indicative of extensive halogen—metal exchange. Thus with various ratios of reactants, a wide range of products are obtained, including platinum hydride complexes, which are most probably formed by the hydrolysis of platinum–lithium intermediates—

7

$(Et_3P)_2PtI_2 + Ph_3GeLi$

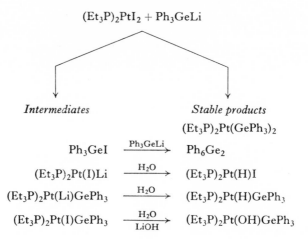

Intermediates		*Stable products*
		$(Et_3P)_2Pt(GePh_3)_2$
Ph_3GeI	$\xrightarrow{Ph_3GeLi}$	Ph_6Ge_2
$(Et_3P)_2Pt(I)Li$	$\xrightarrow{H_2O}$	$(Et_3P)_2Pt(H)I$
$(Et_3P)_2Pt(Li)GePh_3$	$\xrightarrow{H_2O}$	$(Et_3P)_2Pt(H)GePh_3$
$(Et_3P)_2Pt(I)GePh_3$	$\xrightarrow[LiOH]{H_2O}$	$(Et_3P)_2Pt(OH)GePh_3$

The second main preparative method using bistrimethylgermylmercury is probably a general reaction for forming R_3Ge—Pt bonds which has also been used to form trimethylsilyl–platinum complexes—

$$cis\text{-}(Et_3P)_2PtCl_2 + (Me_3Ge)_2Hg \rightarrow trans\text{-}(Et_3P)_2Pt(Cl)GeMe_3 + Me_3GeCl + Hg$$

The reaction is essentially quantitative in boiling benzene if equimolar ratios of reactants are used, but with an excess of the germylmercury compound, coloured intractable oils are formed, rather than the bistrimethylgermylplatinum complex. As will be discussed later, the trimethylgermyl group is probably the most *trans* activating of all ligands, and if two such groups are mutually *trans*, the resulting complex might be unstable, especially in solution.

A further method that could well be developed into the most simple and satisfactory technique for establishing germanium—platinum bonds is illustrated by the reaction—

$$trans\text{-}(Et_3P)_2Pt(H)Cl + Me_3GeH \rightleftharpoons trans\text{-}(Et_3P)_2Pt(Cl)GeMe_3 + H_2$$

The reaction is slow using equimolar ratios of reactants; under one set of conditions some 12% of hydrogen was formed after 20 days at $20°C$. Although attempts to prepare a bistrimethylgermyl complex have been unsuccessful, a mixed trimethylgermyl–triphenylgermyl complex has been prepared by the reaction—

$$trans\text{-}(Et_3P)_2Pt(Cl)GeMe_3 + Ph_3GeLi \rightarrow trans\text{-}(Et_3P)_2Pt(GePh_3)GeMe_3$$

Hexaco-ordinate trichlorogermylplatinum complexes, analogous to the more thoroughly studied tin compounds, have been isolated from reactions between trichlorogermane and platinum(II) halides (Wittle and Urry, 1968)—

$$H_2PtCl_4 \begin{array}{c} \xrightarrow[Me_4NCl]{HGeCl_3} (Me_4N)_2[HPt(GeCl_3)_5] \\[2em] \xrightarrow[Ph_3P]{HGeCl_3} (Ph_3P)_2Pt(Cl)_2(GeCl_3)_2 \end{array}$$

The reactions of germanium–platinum complexes that have been examined are mostly of two types resulting either in retention or cleavage of the metal—metal bond.

A wide variety of both neutral and anionic ligand exchange reactions have been reported that do not involve cleavage of the germanium—platinum bond. Some illustrative examples are given below; but rather surprisingly neither bipyridyl nor triphenylphosphine will substitute triethylphosphine.

$$(Et_3P)_2Pt(GePh_3)_2 \begin{array}{c} \xrightarrow{[Ph_2PCH_2]_2} \quad \begin{bmatrix} Ph_2 \\ P \\ \diagdown \\ Pt \diagdown \begin{array}{c} GePh_3 \\ GePh_3 \end{array} \\ P \\ Ph_2 \end{bmatrix} + 2Et_3P \quad \textbf{(12)} \\[3em] \xrightarrow{2KCN} K_2[(CN)_2Pt(GePh_3)_2] + 2Et_3P \quad \textbf{(13)} \end{array}$$

$$(Et_3P)_2Pt(OH)GePh_3 \underset{H_2O}{\overset{ROH}{\rightleftarrows}} (Et_3P)_2Pt(OR)GePh_3 \quad \textbf{(14)}$$

$$trans\text{-}(Et_3P)_2Pt(Cl)GeMe_3 \begin{array}{c} \xrightarrow{MX} trans\text{-}(Et_3P)_2Pt(X)GeMe_3 \\ (X = Br,\ I,\ SCN) \\[3em] \xrightarrow{(Ph_2PCH_2)_2} \begin{bmatrix} Ph_2 \\ P \\ \diagdown \\ Pt \diagdown \begin{array}{c} PEt_3 \\ GeMe_3 \end{array} \\ P \\ Ph_2 \end{bmatrix}^+ Cl^- \\ \textbf{(15)} \end{array}$$

Formation of the cyanide complex (13) appears to lead to an enhanced thermal stability, as discussed in connection with palladium. The trimethyl-germylplatinum chloride complex gives both *cis* and *trans* monocyanides with 1 mole of potassium cyanide, but with excess complete decomposition to $K_2Pt(CN)_4$ occurs—

7*

$$\textit{trans-}(Et_3P)_2Pt(Cl)GeMe_3 + KCN \longrightarrow \underset{Et_3P}{\overset{Et_3P}{>}}Pt\underset{GeMe_3}{\overset{CN}{<}} + \underset{NC}{\overset{Et_3P}{>}}Pt\underset{PEt_3}{\overset{GeMe_3}{<}}$$

$$\nu(CN),\ 2125\ cm^{-1} \qquad \nu(CN),\ 2119\ cm^{-1}$$

Potassium thiocyanate does not undergo a similar reaction.

Isolation of stable alkoxyplatinum complexes such as **(14)** is unexpected, in view of the instability of other alkoxyplatinum compounds, which decompose spontaneously into platinum hydride complexes and aldehyde, e.g.—

$$(R_3P)_2PtCl_2 + EtOH/NaOH \longrightarrow (R_3P)_2Pt(OEt)Cl$$

$$\textit{trans-}(R_3P)_2Pt(H)Cl + CH_3CHO$$

Perhaps the most remarkable ligand-exchange reaction is that leading to the stable cationic complex **(15)**, rather than a neutral *cis* complex analogous to **(12)**. The high *trans* effect of the Me_3Ge group so weakens the platinum—chlorine bond that expulsion of a stable chloride ion predominates. A similar reaction on the germylplatinum hydride, $(Et_3P)_2Pt(H)GePh_3$, results in cleavage of the germanium—platinum bond and formation of a platinum(o) complex—

$$2\left[\overset{\overset{Ph_2}{-P}}{\underset{\underset{Ph_2}{-P}}{}}\right] + (Et_3P)_2Pt(H)GePh_3$$

$$2Et_3P + \left[\overset{\overset{Ph_2\ Ph_2}{-P\ \ P-}}{\underset{\underset{Ph_2\ Ph_2}{-P\ \ P-}}{Pt}}\right] + Ph_3GeH$$

The bistriphenylgermylplatinum complexes are considerably more thermally stable than similar palladium complexes, though the decomposition products are similar. Above 150°C, some decomposition takes place, and at 220°C products resulting from free-radical cleavage reactions have been isolated—

$$(Et_3P)_2Pt(GePh_3)_2$$

$$\downarrow 220°C$$

$$Et_3P + C_2H_4 + Ph_4Ge + Ph_6Ge_2 + C_6H_6 + \text{"black residue"}$$

The "black residue", containing all of the platinum is soluble in benzene, but on further pyrolysis it loses more Ph_4Ge and Ph_6Ge_2, and is then no longer soluble, although it still retains some 20% of carbon as both phenyl and ethyl groups. Lower molecular-weight complexes, such as *trans*-$(Et_3P)_2Pt(Cl)GeMe_3$, may be sublimed *in vacuo* at 100°C without decomposition.

Reactions which cleave the germanium—platinum bond. All germanium–platinum (and silicon–platinum) compounds that have been examined are cleaved by hydrogen chloride in ethereal or benzene solution. These reactions are considered to proceed via octahedral platinum(IV) intermediates formed by the addition of hydrogen chloride to platinum, with subsequent elimination of the germane or chlorogermane. This general mechanism accounts satisfactorily for the variety of stable cleavage products. In one case, an adduct has been isolated; when dissolved in methanol or acetone at 20° it rapidly decomposes with cleavage of the germanium—platinum bond. This adduct, originally formulated as the platinum(VI) complex (16) is more probably a salt having the structure [(chelate)-$Pt(PEt_3)GeMe_3]^+[HCl_2]^-$ (Hooton, 1968, private communication).

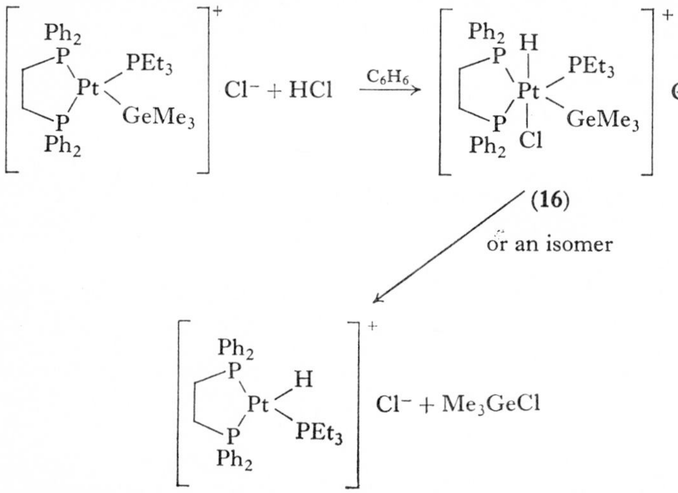

(16)

or an isomer

There are several other examples in which platinum–HCl adducts have proved sufficiently stable to be isolated (Chatt and Shaw, 1962; Cariati *et al.*, 1966)—

$$(R_3P)_2Pt(H)Cl + HCl \rightarrow (R_3P)_2Pt(H)_2Cl_2$$
$$(Ph_3P)_4Pt + 2HCl \rightarrow (Ph_3P)_2Pt(H)_2Cl_2 + 2Ph_3P$$

trans-$(Et_3P)_2Pt(Cl)GeMe_3$ is cleaved quantitatively by 1 mole of hydrogen chloride, according to the equation—

$$trans\text{-}(Et_3P)_2Pt(Cl)GeMe_3 + HCl \rightarrow trans\text{-}(Et_3P)_2(H)Cl + Me_3GeCl$$

Hydrogen chloride cleavage of the unsymmetrical complexes *trans*-$(Et_3P)_2Pt(GeMe_3)GePh_3$ and *cis*-$(Et_3P)_2Pt(Ph)GeMe_3$ is highly selective, the Me$_3$Ge group being cleaved almost entirely as trimethylgermane in each case—

$$(Et_3P)_2Pt(R)GeMe_3 + HCl \rightarrow (Et_3P)_2Pt(R)Cl + Me_3GeH \quad (R = Ph, Ph_3Ge)$$

The selectivity of these reactions may be no more than a kinetic effect, although the stereochemistry of the hexaco-ordinate intermediate could determine whether a germane (R$_3$GeH) or chlorogermane (R$_3$GeCl) is eliminated. The available evidence suggests that with an excess of hydrogen chloride, *cis* complexes give the platinum dichloride (R$_3$P)$_2$PtCl$_2$ whereas *trans* complexes are converted into the platinum hydrido chloride. Thus the octahedral intermediate (17) in the case of a *cis* complex has both hydrogen and chlorine *cis* to the Ph$_3$Ge group and selectively eliminates triphenylgermane—

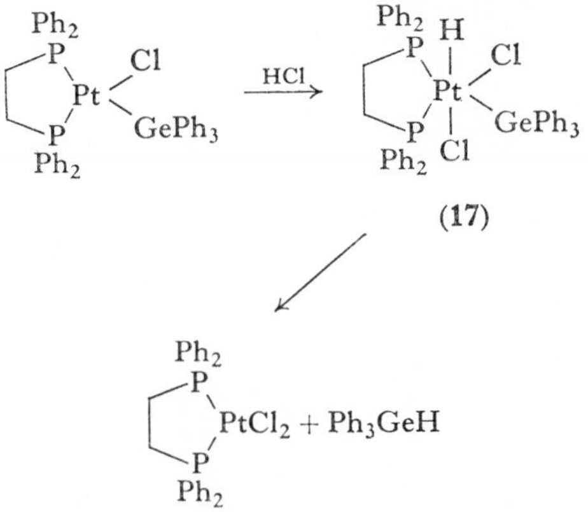

(17)

However in *trans* complexes *if* the octahedral intermediate has hydrogen *trans* to germanium, as in (18), then *cis* elimination of Me$_3$GeCl must occur—

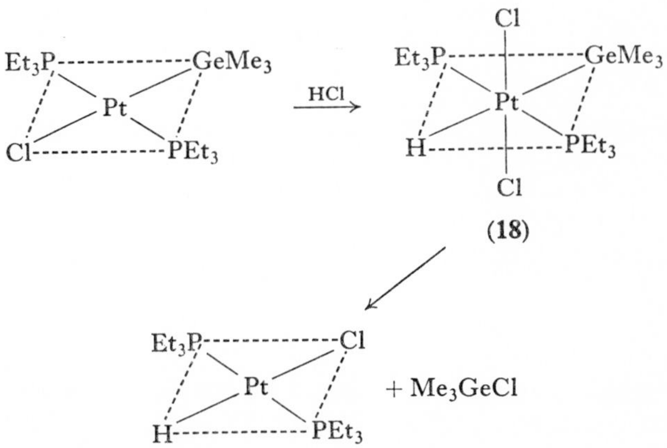

(18)

Dipole moment evidence confirms that the complex $(Et_3P)_2Pt(GePh_3)_2$ exists in solution as an equilibrium mixture of *cis* and *trans* forms and the final products of its cleavage by hydrogen chloride are consistent with the above interpretation of the reaction—

$$cis\text{-, } trans\text{-}(Et_3P)_2Pt(GePh_3)_2$$

$$\downarrow HCl$$

$$cis\text{-}(Et_3P)_2Pt(Cl)GePh_3 + trans\text{-}(Et_3P)_2Pt(Cl)GePh_3 + Ph_3GeH$$

$$\downarrow HCl \qquad\qquad\qquad \downarrow HCl$$

$$cis\text{- or } trans\text{-}(Et_3P)_2PtCl_2 + Ph_3GeH \qquad trans\text{-}(Et_3P)_2Pt(H)Cl + Ph_3GeCl$$

The triphenylgermylplatinum complexes, $(R_3P)_2Pt(GePh_3)_2$ are stable to water and alcoholic potassium hydroxide, whereas the trimethylgermyl group in $(Et_3P)_2Pt(Cl)GeMe_3$ is slowly hydrolysed by aqueous diglyme—

$$(Et_3P)_2Pt(Cl)GeMe_3 + \tfrac{1}{2}H_2O \rightarrow (Et_3P)_2Pt(H)Cl + \tfrac{1}{2}(Me_3Ge)_2O$$
$$\text{Half-life} \sim 215 \text{ hr at } 18°C$$

By contrast, the trimethylsilyl analogue is completely hydrolysed within a few minutes under the same conditions. The difference possibly suggests primary attack by water at the silicon or germanium atom, but with the complex cation (**19**) the rate of hydrolysis is strikingly reduced (30% after 50 days when $M = SiMe_3$ and 20% after 100 days when $M = GeMe_3$). Moreover the rate of hydrolysis is even lower in the presence of a base—

$$\left[\begin{array}{c} Ph_2 \\ P \\ \diagup \diagdown \\ Pt \diagup^{MMe_3} \\ P \diagdown PEt_3 \\ Ph_2 \end{array} \right]^+ Cl^- + \tfrac{1}{2}H_2O \longrightarrow \left[\begin{array}{c} Ph_2 \\ P \\ \diagup \diagdown H \\ Pt \diagdown PEt_3 \\ P \\ Ph_2 \end{array} \right]^+ Cl^- + \tfrac{1}{2}(Me_3M)_2O$$

(**19**)

Since in the complex cation the positive charge probably resides largely on platinum, these observations on relative rates of hydrolysis are most compatible with proton attack at platinum as the rate-determining step.

The hydrogenolysis of germanium–platinum and silicon–platinum complexes is a reversible process, requiring a low activation energy and, like the hydrogen chloride cleavage reactions, it probably occurs via an octahedral addition complex. Reaction-rate measurements on the hydrogenolysis of $(Et_3P)_2Pt(GePh_3)_2$ in toluene solution indicate an activation energy of about 9 kcal. mole^{-1}. At 0°C and 0·25 atm hydrogen, the reaction proceeds to 55% after 24 hr, according to the equation—

$$(Et_3P)_2Pt(GePh_3)_2 + H_2 \rightarrow (Et_3P)_2Pt(H)GePh_3 + Ph_3GeH$$

Trimethylgermyl (and silyl) groups bonded to platinum are even more readily cleaved by hydrogen, both from neutral and from cationic complexes, and the reaction is extremely selective for the complexes $(Et_3P)_2Pt(GePh_3)GeMe_3$ and $(Et_3P)_2Pt(Ph)GeMe_3$—

$$trans\text{-}(Et_3P)_2Pt(Cl)GeMe_3 + H_2 \rightleftharpoons trans\text{-}(Et_3P)_2Pt(H)Cl + Me_3GeH$$

$$\left[\begin{array}{c} Ph_2 \\ \overset{P}{\diagdown} \hspace{-0.5em} \underset{P}{\diagup} Pt \overset{PEt_3}{\underset{GeMe_3}{\diagdown}} \\ Ph_2 \end{array}\right]^+ Cl^- + H_2 \longrightarrow \left[\begin{array}{c} Ph_2 \\ \overset{P}{\diagdown} \hspace{-0.5em} \underset{P}{\diagup} Pt \overset{PEt_3}{\underset{H}{\diagdown}} \\ Ph_2 \end{array}\right]^+ Cl^- + Me_3GeH$$

$$trans\text{-}(Et_3P)_2Pt(GeMe_3)GePh_3 + H_2 \longrightarrow trans\text{-}(Et_3P)_2Pt(H)GePh_3 + Me_3GeH$$

The last of these reactions proceeds to completion within a few minutes at 20°C and 1 atm hydrogen. One aspect that is probably relevant to the mechanism is that *cis* platinum complexes and palladium compounds are less readily hydrogenated. In the following examples, hydrogenolysis proceeds rapidly only with fairly high (100 atm) pressure of hydrogen—

$$(Et_3P)_2Pd(GePh_3)_2 + H_2 \xrightarrow[20°C]{100\ Atm} (Et_3P)_2Pd(H)GePh_3 + Ph_3GeH$$

$$\begin{array}{c} Ph_2 \\ \overset{P}{\diagdown} \hspace{-0.5em} \underset{P}{\diagup} Pt \overset{GePh_3}{\underset{GePh_3}{\diagdown}} \\ Ph_2 \end{array} + H_2 \xrightarrow[50°C]{100\ Atm} \begin{array}{c} Ph_2 \\ \overset{P}{\diagdown} \hspace{-0.5em} \underset{P}{\diagup} Pt \overset{H}{\underset{GePh_3}{\diagdown}} \\ Ph_2 \end{array} + Ph_3GeH$$

$$cis\text{-}(Et_3P)_2Pt(Ph)GeMe_3 + H_2 \longrightarrow$$
$$cis\text{-}(Et_3P)_2Pt(H)GeMe_3 + C_6H_6 + \text{trace } Me_3GeH$$

The low activation energy of all these homogeneous hydrogenolysis reactions suggests an addition followed by a *cis* elimination mechanism. Although octahedral intermediates have not been isolated or detected spectroscopically, the reversibility of the reactions is most readily explained in these terms. Moreover, other transition-metal complexes are known to add hydrogen reversibly, and, for the iridium complex at least, *cis* addition has been established (Vaska, 1966)—

$$(R_3P)_2Ir(CO)Cl + H_2 \rightleftharpoons (R_3P)_2(CO)Ir(H)_2Cl$$

In the case of platinum there are four possible structures for an octahedral intermediate of the type $(R_3P)_2Pt(H)_2ab$ if one assumes *cis* addition of hydrogen. Three of these (**20–22**) have the phosphine groups *cis* to each other, whereas the fourth structure (**23**) is most probably the labile intermediate for initially *trans* complexes—

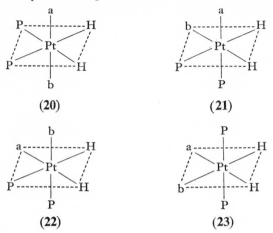

(**20**) (**21**)

(**22**) (**23**)

The stability of these germylplatinum hydride complexes varies considerably. $(Et_3P)_2Pt(H)GePh_3$ is stable to air, water, ethanol and ethanolic potassium hydroxide, but when heated in xylene some decomposition occurs, with deposition of platinum. The trimethylgermylplatinum hydride *cis*-$(Et_3P)_2Pt(H)GeMe_3$ is much less stable and has only been obtained as an oil. Both complexes have very low platinum–hydrogen stretching frequencies and perhaps the difference in their stability is due to the high crystal stabilization energy for the triphenylgermyl complex.

Lithium aluminium hydride reduction of $(Pr^n_3P)_2Pt(GePh_3)_2$ results in cleavage of both germanium—platinum bonds, forming triphenylgermane and an insoluble platinum-containing product that on hydrolysis decomposes to platinum and hydrogen.

The metal—metal bonds in most germanium–transition metal complexes so far reported are readily cleaved by 1,2-dibromoethane; the reactions with platinum and palladium complexes are essentially quantitative at room temperature—

$$(Et_3P)_2Pt(GePh_3)_2 + 2C_2H_4Br_2 \rightarrow \textit{trans-}(Et_3P)_2PtBr_2 + 2Ph_3GeBr + 2C_2H_4$$

In this example, the reaction rate decreases significantly after 1 mole of 1,2-dibromoethane has reacted.

The possible reaction mechanism is discussed in more detail in connection with germanium–gold complexes.

Carbon tetrachloride is sufficiently reactive to cleave the germanium—platinum bond, but the fate of the CCl_3 group has not been determined—

$$(R_3P)_2Pt(GePh_3)_2 + CCl_4 \rightarrow 2Ph_3GeCl + (R_3P)_2PtCl_2$$

Methyl iodide will cleave the germanium—platinum bond at 110°C, but this reaction evidently results in radical formation, since methane and ethane are formed together with Ph_3GeMe and Ph_3GeI.

Iodine reacts rapidly at room temperature with bistriphenylgermyl-platinum complexes, and the intermediate triphenylgermylplatinum iodide is more rapidly cleaved than the original complex—

$$(R_3P)_2Pt(GePh_3)_2 \xrightarrow{I_2} (R_3P)_2PtI_2 + 2Ph_3GeI$$

In this reaction there is no evidence for the cleavage of germanium—phenyl bonds, in contrast to the behaviour of so many triphenyltin–transition metal complexes.

Magnesium iodide brings about the cleavage of germanium—platinum bonds to give various products that require the formation of platinum–Grignard reagents as intermediates. If the reaction is analogous to hydrogen chloride cleavage, then the diversity of products may be partly accounted for by the occurrence of *cis–trans* forms of the initial complex—

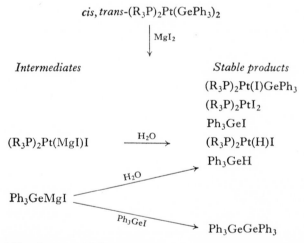

$$cis, trans\text{-}(R_3P)_2Pt(GePh_3)_2$$

Nucleophilic cleavage of the germanium—platinum bond appears to be a slow, and possibly reversible, reaction—

$$(R_3P)_2Pt(GePh_3)_2 + 2PhLi \rightleftharpoons (R_3P)_2PtPh_2 + 2Ph_3GeLi$$

It shows no indications of paralleling the analogous cleavage reaction of gold–germanium complexes.

Both i.r. and p.m.r. spectroscopy have been used to investigate the stereochemistry of these complexes, and in addition i.r. spectra can give information on the relative *trans* effects of ligands bonded to platinum.

Trans complexes usually show one medium to strong band [$\nu_{asym.}$ (Pt–P)] in the range 418—406 cm^{-1}, whereas *cis* complexes show a medium to strong doublet, $\nu_{sym.}$ and $\nu_{asym.}$ (Pt–P), near 440 and 425 cm^{-1}. In the case of germanium–platinum complexes, this relationship can be used with reasonable confidence in assigning the stereochemistry. One exception seems to be the colourless complex *cis*-(Et$_3$P)$_2$Pt(GePh$_3$)$_2$, which shows only one band at 413 cm^{-1} as a nujol mull, whereas the solid cyanide, (Et$_3$P)$_2$Pt(CN)GeMe$_3$, shows three bands in the region of ν(Pt–P), and hence an unambiguous assignment is not possible. In the *trans* complexes (Et$_3$P)$_2$Pt(Cl)SiMe$_3$ and (Et$_3$P)$_2$Pt(Cl)GeMe$_3$, the platinum–chlorine stretching frequency occurs at 238 and 235 cm^{-1}, respectively, which is considerably lower than the usual values (269–340 cm^{-1}), and this undoubtedly reflects the extremely high *trans* effect of Me$_3$Ge and Me$_3$Si groups bonded to platinum. The two hydridoplatinum complexes (Et$_3$P)$_2$Pt(H)GePh$_3$ and (Pr$^n_3$P)$_2$Pt(H)GePh$_3$ are of interest in that the platinum–hydrogen stretching frequency occurs at 2051 and 1957 cm^{-1}, respectively. This suggests that the tri-n-propylphosphine complex has a *trans* structure, since the *trans* effect of Ph$_3$Ge is probably greater than that of R$_3$P, and hence shifts ν(Pt–H) to lower frequency.

The stereochemistry of trimethylgermyl- (and trimethylsilyl-) platinum complexes in solution can be determined unambiguously from the spectrum of the Me$_3$Ge or Me$_3$Si protons. In the complex *trans*-(Et$_3$P)$_2$Pt(Cl)GeMe$_3$ (Fig. 8) the Me$_3$Ge resonance consists of a single sharp line (9·42τ) flanked by two satellites produced by coupling with the ^{195}Pt nucleus ($I = \frac{1}{2}$, 34% abundant). The relative intensities (1:4:1) are as expected, and the coupling constant $J(^{195}$Pt–Ge–C–^{1}H) = 20 c/s. The methyl protons of the triethylphosphine groups form a 5-line system (9·01τ) having the relative intensities 1:4:6:4:1 and a line separation of 7·75 c/s. This may be described in terms of splitting of the CH$_3$ resonance by the CH$_2$ protons to give a 1:2:1 triplet which is further split, by coupling to the two equivalent *trans* ^{31}P nuclei, with an apparent coupling constant $J(^{31}$P–^{1}H) equal to J(CH$_3$–CH$_2$) and equal to the observed line separation. The CH$_2$ resonance of the triethylphosphine groups (8·1τ) consists of at least 13 lines with a mean separation of about 4 c/s. This is clearly the result of coupling with the CH$_3$ protons of the ethyl groups, the two ^{31}P nuclei and the ^{195}Pt nucleus.

The p.m.r. spectra of trimethylgermyl platinum complexes in which the two phosphine groups are mutually *cis* are profoundly different. For example, in *cis*-(Et$_3$P)$_2$Pt(Ph)GeMe$_3$ (Fig. 8) the ethyl proton resonances are far more complex and show little symmetry. Each line of the Me$_3$Ge resonance (9·67τ) appears as a symmetrical doublet, thus producing a 6-line spectrum, having relative intensities of 1:1:4:4:1:1, the line separation of the doublets being 1·4 c/s. The coupling constant to platinum [$J(^{195}$Pt–Ge–C–^{1}H)] is 15·6 c/s, considerably lower than in most *trans* isomers. This additional splitting of the Me_3Ge resonances is attributable to long-range coupling to the *trans* ^{31}P nucleus [$J(^{31}P-^{195}Pt-Ge-C-^1H)$], and is analogous to the coupling observed between ^{31}P nuclei in *trans* complexes.

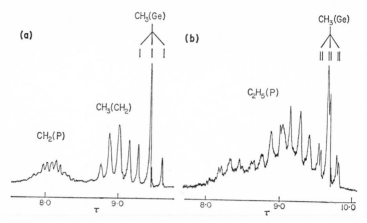

FIG. 8. ^1H n.m.r. spectra of *cis*- and *trans*-trimethylgermylplatinum(II) complexes: (a), *trans*; (b), *cis*. (Reproduced from *J. chem. Soc.*, by courtesy of The Chemical Society.)

Of particular interest is the complex $(Et_3P)_2Pt(GePh_3)GeMe_3$, for which the i.r. evidence suggests a *trans* configuration in the solid state. Its p.m.r. spectrum is complex, and coupling between the Me_3Ge protons and the ^{195}Pt nucleus is not observed. These complexities probably result from an equilibrium between *cis* and *trans* forms in solution. The coupling constants, $J(^{195}Pt-Ge-C-^1H)$, for a series of complexes are shown in Table XV, from

TABLE XV

Platinum–proton coupling constants in Me_3GePt **complexes**

Complex	$J(^1H-C-Ge-^{195}Pt)$, c/s	*trans* ligand
trans-$(Et_3P)_2Pt(Cl)GeMe_3$	20·0	Cl
trans-$(Et_3P)_2Pt(Br)GeMe_3$	20·0	Br
trans-$(Et_3P)_2Pt(I)GeMe_3$	19·9	I
trans-$(Et_3P)_2Pt(NCS)GeMe_3$	17·7	NCS
cis-$(Et_3P)_2Pt(CN)GeMe_3$	16·8	Et_3P
cis-$(Et_3P)_2Pt(Ph)GeMe_3$	15·6	Et_3P
cis-$[Ph_2PCH_2CH_2PPh_2Pt(PEt_3)GeMe_3]^+Cl^-$	12·8	Ph_2PCH_2-
trans-$(Et_3P)_2Pt(CN)GeMe_3$	11·8	CN

which it is seen that the values decrease as the *trans* effect of the group *trans* to Me_3Ge increases. This trend is in line with a weakening of the germanium —platinum bond by ligands of high *trans* effect.

H. Copper, silver and gold

Triphenylgermyl-lithium reacts with monotriphenylphosphine complexes of copper, silver and gold in their +1 oxidation state to form sparingly soluble metal—metal bonded compounds (Glockling and Hooton, 1962a; Glockling and Wilbey, 1968)—

$$Ph_3PMCl + Ph_3GeLi \rightarrow Ph_3PMGePh_3 + LiCl$$

The copper(I) and silver(I) compounds, which are sensitive to aerial oxidation and hydrolysis, are more thermally stable and less reactive if excess triphenylphosphine is present when tetraco-ordinate complexes are obtained—

$$(Ph_3P)_3CuI + Ph_3GeLi \rightarrow (Ph_3P)_3CuGePh_3 + LiI$$

Germanium–gold compounds are more readily investigated and the triphenylphosphine complex, $Ph_3PAuGePh_3$ is stable in air over at least 5 years, and is unaffected by water. In peroxidized ethereal solvents it decomposes with deposition of gold. The trimethylphosphine analogue, which is monomeric in benzene, is considerably less stable thermally. The complex $Et_3PAuGePh_3$ is sufficiently volatile and stable to produce an abundant molecular ion under electron impact. A similar trimethylgermylgold complex has been isolated in low yield from the reaction—

$$Ph_3PAuCl + (Me_3Ge)_2Hg \rightarrow Ph_3PAuGeMe_3 + Hg + Me_3GeCl$$

It is less thermally stable than the triphenylgermyl complex, and oxidizes in air when in solution. The formation of dico-ordinate complexes is a characteristic feature of gold chemistry, but there is evidence that the co-ordination number can increase as in the chelating phosphine complex—

$$Ph_3PAuGePh_3 + (Ph_2PCH_2)_2 \longrightarrow Ph_3P + \overset{\displaystyle Ph_2}{\underset{\displaystyle Ph_2}{\left[\begin{array}{c} P \\ P \end{array}\right.}} \!\!\!AuGePh_3$$

In its reactions with polar reagents, the germanium—gold bond shows the polarity $\overset{\delta+}{Au}—\overset{\delta-}{Ge}$. Dibromoethane reacts rapidly and quantitatively at room temperature—

$$Ph_3PAuGePh_3 + C_2H_4Br_2 \rightarrow Ph_3PAuBr + Ph_3GeBr + C_2H_4$$

This reaction, which has already been referred to, seems to be quite general for germanium—transition metal complexes. Although there is no mechanistic information available, it is tempting to suggest a 4-centred transition intermediate, and it would be of interest to determine whether *cis* and *trans* olefins are formed stereospecifically from other vicinal dibromides. Ethylene dichloride reacts in the same way, but much more slowly at room temperature. At higher temperatures, a competing reaction occurs in which gold is formed together with a phosphonium salt.

$$Ph_3PAuGePh_3 + C_2H_4Cl_2 \longrightarrow Ph_3PAuCl + Ph_3GeCl + C_2H_4$$

$$\downarrow Ph_3PAuGePh_3$$

$$Au + Ph_3GeCl + [Ph_3PCH_2CH_2PPh_3]Cl_2$$

The germanium—gold bond is unreactive to hydrogen at 1 atm, but at high pressure it is cleaved, and gold is deposited, possibly through an intermediate unstable gold hydride—

$$Ph_3PAuGePh_3 + H_2 \rightarrow Ph_3P + Ph_3GeH + Au$$

The germanium—gold bond is broken by reaction with methanolic potassium cyanide—

$$Ph_3PAuGePh_3 \xrightarrow[\text{MeOH}]{\text{KCN}} K[Ph_3GeAuCN] + Ph_3P$$

$$\downarrow \text{MeOH}$$

$$KAu(CN)_2 \xleftarrow{\text{KCN}} K[MeOAuCN] + Ph_3GeH$$

Other reagents which cleave the germanium—gold bond are summarized below—

$$Ph_3PAuGePh_3 \begin{cases} \xrightarrow{\text{HCl}} Ph_3PAuCl + Ph_3GeH \\ \xrightarrow{\text{CCl}_4} Ph_3PAuCl + Ph_3GeCl \\ \xrightarrow{\text{HgCl}_2} Ph_3PAuCl + Ph_3GeCl + Hg \\ \xrightarrow{\text{MeI}} Ph_3PAuI + Ph_3GeMe + Au + C_2H_6 + CH_4 \end{cases}$$

Cleavage by mercuric chloride is a reaction common to other metal—metal bonded compounds, and may proceed via an unstable germylmercury intermediate, $Ph_3GeHgCl$, or a 5-centred transition state.

The reaction with stannic chloride is of interest, in that the germanium—gold bond is cleaved and a tin—gold bond formed—

$$Ph_3PAuGePh_3 + SnCl_4 \nearrow Ph_3GeCl + Ph_3PAuCl + SnCl_2$$
$$\searrow Ph_3GeCl + (Ph_3P)_2AuSnCl_3 + Au$$

Similar tin–gold complexes have been obtained by the reaction (Dilts and Johnson, 1966)—

$$Ph_3PMCl + SnCl_2 + 2Ph_3P \rightarrow (Ph_3P)_3MSnCl_3$$
$$(M = Cu, Ag, Au)$$

Phenyl-lithium cleaves the germanium—gold bond to give an unstable anionic complex that has been isolated as the tetra-alkylammonium salt—

$$Ph_3PAuGePh_3 + PhLi \longrightarrow Ph_3GeLi + Ph_3PAuPh$$
$$\downarrow Ph_3PAuGePh_3$$
$$Et_4N[(Ph_3Ge)_2Au] \xleftarrow{Et_4NCl} Li[(Ph_3Ge)_2Au] + Ph_3P$$

The rate of attack of triphenylgermyl-lithium on the germanium–gold complex is greater than its rate of cleavage by phenyl-lithium. This is one of the few examples of a stable metal—metal bonded complex that does not contain strongly π-bonding ligands. Cleavage of the germanium—gold bond by magnesium bromide seems to be an equilibrium reaction—

$$Ph_3PAuGePh_3 + MgBr_2 \rightleftharpoons Ph_3PAuBr + Ph_3GeMgBr$$
$$\swarrow H_2O$$
$$Ph_3GeH$$

REFERENCES

Abel, E. W., and Armitage, D. A. (1967). *Adv. organometal. Chem.*, **5**, 1.
Abel, E. W., and Brady, D. B. (1968). *J. organometal. Chem.*, **11**, 145.
Abel, E. W., Armitage, D. A., and Brady, D. B. (1966). *J. organometal. Chem.*, **5**, 130.
Adley, A. D., Gilson, D. F. R., and Onyszchuk, M. (1968). *Chem. Commun.*, 813.
Adveeva, V. I., Burlachenko, G. S., Baukov, Yu. I., and Lutsenko, I. F. (1966). *Zh. obshch. Khim. (Eng. Transl.)*, **36**, 1676.
Agolini, F., Klemenko, S., Csizmadia, I. G., and Yates, K. (1968). *Spectrochim. Acta*, **24A**, 169.
Allison, E. R., and Mueller, J. H. (1932). *J. Am. chem. Soc.*, **54**, 2833.
Allred, A. L. (1961). *J. inorg. nucl. Chem.*, **17**, 215.
Allred, A. L., and Nicholson, D. A. (1965). *Inorg. Chem.*, **4**, 1747, 1751.
Allred, A. L., and Rochow, E. G. (1958). *J. inorg. nucl. Chem.*, **5**, 269.
Almenningen, A., Bastiansen, O., Ewing, V., Hedberg, K., and Traetteberg, M. (1963). *Acta chem. scand.*, **17**, 2455.
Amberger, E. (1959). *Angew. Chem.*, **71**, 372.
Amberger, E., and Boeters, H. (1961). *Angew. Chem.*, **73**, 114.
Amberger, E., and Boeters, H. (1963). *Angew. Chem., Int. Ed.*, **2**, 686.
Amberger, E., and Muehlhofer, E. (1968). *J. organometal. Chem.*, **12**, 55
Amberger, E., and Salazar, R. W. (1967). *J. organometal. Chem.*, **8**, 111.
Amberger, E., Stoeger, W., and Grossich, H. R. (1966). *Angew. Chem., Int. Ed.*, **5**, 522.
Anderko, K., and Schubert, K. (1953). *Z. Metallk.* **44**, 307.
Anderson, D. G., Chipperfield, J. R., and Webster, D. E. (1968). *J. organometal. Chem.*, **12**, 323.
Anderson, H. H. (1949). *J. Am. chem. Soc.*, **71**, 1799.
Anderson, H. H. (1950). *J. Am. chem. Soc.*, **72**, 194, 2089.
Anderson, H. H. (1951a). *J. Am. chem. Soc.*, **73**, 5439, 5440.
Anderson, H. H. (1951b). *J. Am. chem. Soc.*, **73**, 5798, 5800.
Anderson, H. H. (1952a). *J. Am. chem. Soc.*, **74**, 1421.
Anderson, H. H. (1952b). *J. Am. chem. Soc.*, **74**, 2370, 2371.
Anderson, H. H. (1953). *J. Am. chem. Soc.*, **75**, 814.
Anderson, H. H. (1955). *J. org. Chem.*, **20**, 536.
Anderson, H. H. (1956a). *J. Am. chem. Soc.*, **78**, 1692.
Anderson, H. H. (1956b). *J. org. Chem.*, **21**, 869.
Anderson, H. H. (1957). *J. Am. chem. Soc.*, **79**, 326.
Anderson, H. H. (1960). *J. Am. chem. Soc.*, **82**, 3016.
Anderson, H. H. (1961). *J. Am. chem. Soc.*, **83**, 547.
Anderson, H. H. (1964). *Inorg. Chem.*, **3**, 910.
Andrews, T. D., and Phillips, C. S. G. (1966). *J. chem. Soc. A*, 46.
Andrieux, J. L., and Andrieux, M. J. B. (1955). *C.r. hebd. Séanc. Acad. Sci. Paris*, **240**, 2104.
Armer, B., and Schmidbaur, H. (1967). *Chem. Ber.*, **100**, 1521.

Aronson, J. R., and Durig, J. R. (1964). *Spectrochim. Acta*, **20**, 219.

Arvedson, P., and Larsen, E. M. (1966). *Inorg. Synth.*, **8**, 34.

Avduevskaya, K. A., and Tananaev, I. V. (1965). *Zh. neorg. Khim. (Eng. Transl.)*, **10**, 197.

Avduevskaya, K. A., Tananaev, I. V., and Mironova, V. S. (1964). *Dokl. Akad. Nauk SSSR (Eng. Transl.)*, **157**, 747.

Aylett, B. J., and Campbell, J. M. (1965). *Chem. Commun.*, 159.

Baay, Y. L., and MacDiarmid, A. G. (1967). *Inorg. nucl. Chem. Lett.*, **3**, 159.

Baird, M. C. (1967). *J. inorg. nucl. Chem.*, **29**, 367.

Bartlett, N., and Yu, K. C. (1961). *Can. J. Chem.*, **39**, 80.

Bauer, H., and Burschkies, K. (1932). *Chem. Ber.*, **65**, 956.

Bauer, H., and Burschkies, K. (1934). *Chem. Ber.*, **67**, 1041.

Baukov, Yu. I., and Lutsenko, I. F. (1964). *Zh. obshch. Khim. (Eng. Transl.)*, **34**, 3495.

Baukov, Yu. I., Belavin, I. Yu., and Lutsenko, I. F. (1965a). *Zh. obshch. Khim. (Eng. Transl.)*, **35**, 1096.

Baukov, Yu. I., Burlachenko, G. S., and Lutsenko, I. F. (1965b). *Zh. obshch. Khim. (Eng. Transl.)*, **35**, 1178.

Baukov, Yu., Burlachenko, G. S., Belavin, I. Yu., and Lutsenko, I. F. (1966). *Zh. obshch. Khim. (Eng. Transl.)*, **36**, 158.

Baum, G., Lehn, W. L., and Tamborski, C. (1964). *J. org. Chem.*, **29**, 1264.

Beagley, B., Robiette, A. G., and Sheldrick, G. M. (1967). *Chem. Commun.*, 601.

Beattie, I. R., McQuillan, G. P., Rule, L., and Webster, M. (1963). *J. chem. Soc.*, 1514.

Beattie, I. R., Webster, M., and Chantry, G. W. (1964). *J. chem. Soc.*, 6172.

Beattie, I. R., Gilson, T., Livingstone, K., Fawcett, V., and Ozin, G. A. (1967). *J. chem. Soc. A*, 712.

Beattie, I. R., Gilson, T., and Ozin, G. A. (1968). *J. chem. Soc. A*, 1092.

Beck, W., and Schuierer, E. (1964). *Chem. Ber.*, **97**, 3517.

Bedford, J. A., Bolton, J. R., Carrington, A., and Prince, R. H. (1963). *Trans. Faraday Soc.*, **59**, 53.

Begun, G. M., and Rutenberg, A. C. (1967). *Inorg. Chem.*, **6**, 2212.

Benkeser, R. A., De Boer, C. E., Robinson, R. E., and Sauve, D. M. (1956). *J. Am. chem. Soc.*, **78**, 682.

Billig, E. (1955). *Proc. R. Soc.*, **A229**, 346.

Bills, J. L., and Cotton, F. A. (1964). *J. phys. Chem., Ithaca*, **68**, 806.

Birchall, T., and Jolly, W. L. (1966). *Inorg. Chem.*, **5**, 2177.

Birr, K. H. (1962). *Z. anorg. Chem.*, **315**, 175.

Birr, K. H., and Kraeft, D. (1961). *Z. anorg. Chem.*, **311**, 235.

Bither, T. A., Knoth, W. H., Lindsey, R. V., and Sharkey, W. H. (1958). *J. Am. chem. Soc.*, **80**, 4151.

Bokii, N. K., and Struchkov, Yu. T. (1967). *Zh. strukt. Khim. (Eng. Transl.)*, **8**, 100.

Bolton, J. R., and Carrington, A. (1961). *Molec. Phys.*, **4**, 497.

Booth, R. B., and Kraus, C. A. (1952). *J. Am. chem. Soc.*, **74**, 1415.

Bott, R. W., Eaborn, C., Pande, K. C., and Swaddle, T. W. (1962). *J. chem. Soc.*, 1217.

Bott, R. W., Eaborn, C., and Swaddle, T. W. (1963a). *J. chem. Soc.*, 2342.

Bott, R. W., Eaborn, C., and Hashimoto, T. (1963b). *J. chem. Soc.*, 3906.

Bott, R. W., Eaborn, C., and Walton, D. R. M. (1964a). *J. organometal. Chem.*, **2**, 154.

Bott, R. W., Eaborn, C., and Greasley, P. M. (1964b). *J. chem. Soc.*, 4804.
Bottei, R. S., and Kuzma, L. J. (1968). *J. inorg. nucl. Chem.*, **30**, 415.
Bradley, D. C., Kay, L. J., and Wardlaw, W. (1956). *J. chem. Soc.*, 4916.
Brauer, G. (Ed.) (1963). "Handbook of Preparative Inorganic Chemistry", 2nd Ed., Vol. 1. Academic Press, New York.
Briggs, T. R., and Benedict, W. S. (1930). *J. phys. Chem.*, Ithaca, **34**, 173.
Brinckman, F. E., and Stone, F. G. A. (1959). *J. inorg. nucl. Chem.*, **11**, 24.
Brook, A. G. (1955). *J. Am. chem. Soc.*, **77**, 4827.
Brook, A. G., and Fieldhouse, S. A. (1967). *J. organometal. Chem.*, **10**, 235.
Brook, A. G., and Gilman, H. (1954). *J. Am. chem. Soc.*, **76**, 77.
Brook, A. G., and Peddle, G. J. D. (1963). *J. Am. chem. Soc.*, **85**, 1869, 2338.
Brook, A. G., and Peddle, G. J. D. (1966). *J. organometal. Chem.*, **5**, 106.
Brook, A. G., Quigley, M. A., Peddle, G. J. D., Schwartz, N. V., and Warner, C. M. (1960). *J. Am. chem. Soc.*, **82**, 5102.
Brook, A. G., Pannell, K. H., LeGrow, G. E., and Sheeto, J. J. (1964). *J. organometal. Chem.*, **2**, 491.
Brook, A. G., Duff, J. M., Jones, P. F., and Davis, N. R. (1967). *J. Am. chem. Soc.*, **89**, 431.
Brooks, E. H., and Glockling, F. (1966). *J. chem. Soc. A*, 1241.
Brooks, E. H., and Glockling, F. (1967). *J. chem. Soc. A*, 1030.
Brooks, E. H., Glockling, F., and Hooton, K. A. (1965). *J. chem. Soc.*, 4283.
Brooks, E. H., Cross, R. J., and Glockling, F. (1968). *Inorg. chim. Acta*, **2**, 17.
Brown, M. P., and Rochow, E. G. (1960). *J. Am. chem. Soc.*, **82**, 4166.
Brown, M. P., Cartmell, E., and Fowles, G. W. A. (1960a). *J. chem. Soc.*, 506.
Brown, M. P., Okawara, R., and Rochow, E. G. (1960b). *Spectrochim. Acta*, **16**, 595.
Brown, D. H., Dixon, K. R., Livingston, C. M., Nuttall, R. H., and Sharp, D. W. A. (1967). *J. chem. Soc. A*, 100.
Bulten, E. J., and Noltes, J. G. (1966a). *Tetrahedron Lett.*, **29**, 3471.
Bulten, E. J., and Noltes, J. G. (1966b). *Tetrahedron Lett.*, **36**, 4389.
Bulten, E. J., and Noltes, J. G. (1967). *Tetrahedron Lett.*, **16**, 1443.
Bulten, E. J., and Noltes, J. G. (1968). *J. organometal. Chem.*, **11**, P 19.
Bush, M. A., and Woodward, P. (1967). *J. chem. Soc. A*, 1833.
Busmann, E. (1961). *Z. anorg. Chem.*, **313**, 90.
Calvin, G. and Coates, G. E. (1960). *J. chem. Soc.*, 2008.
Cariati, F., Ugo, R., and Bonati, F. (1966). *Inorg. Chem.*, **5**, 1128.
Carre, F. H., Corriu, R. J. P., and Thomassin, R. B. (1968). *Chem. Commun.* 560.
Carrick, A., and Glockling, F. (1966). *J. chem. Soc. A*, 623.
Carrick, A., and Glockling, F. (1967). *J. chem. Soc. A*, 40.
Carrick, A., and Glockling, F. (1968). *J. chem. Soc. A*, 913.
Cawley, S., and Danyluk, S. S. (1963). *Can. J. Chem.*, **41**, 1850.
Cerf, C., and Delhaye, M. B. (1964). *Bull. Soc. chim. Fr.*, 2818.
Chalk, A. J., and Harrod, J. F. (1965). *J. Am. chem. Soc.*, **87**, 16, 1133.
Chambers, R. D., and Cunningham, J. (1965). *Tetrahedron Lett.*, **28**, 2389.
Chambers, D. B., and Glockling, F. (1968). *J. chem. Soc. A*, 735.
Chambers, D. B., Glockling, F., and Weston, M. (1967). *J. chem. Soc. A*, 1759.
Chambers, D. B., Glockling, F., and Light, J. R. C. (1968). *Q. Rev. chem. Soc.*, **22**, 317.
Chatt, J., and Shaw, B. L. (1962). *J. chem. Soc.*, 5075.
Chipperfield, J. R., and Prince, R. H. (1963). *J. chem. Soc.*, 3567.
Clark, E. R. (1961). *J. inorg. nucl. Chem.*, **21**, 366.

Clark, H. C., and Dixon, K. R. (1967). *Chem. Commun.*, 717.
Clark, H. C., and Willis, C. J. (1962). *J. Am. chem. Soc.*, **84**, 898.
Clero, P. J. (1966). *Bull. Soc. chim. Fr.*, 2455.
Coffey, C. E., Lewis, J., and Nyholm, R. S. (1964). *J. chem. Soc.*, 1741.
Cohen, S. C., and Massey, A. G. (1968). *J. organometal. Chem.*, **12**, 341.
Colin, R., and Drowart, J. (1964). *J. phys. Chem., Ithaca*, **68**, 428.
Corey, E. J., Seebach, D., and Freedman, R. (1967). *J. Am. chem. Soc.*, **89**, 434.
Cotton, J. D., Knox, S. A. R., and Stone, F. G. A. (1967a). *Chem. Commun.*, 965.
Cotton, J. D., Knox, S. A. R., Paul, I., and Stone, F. G. A. (1967b). *J. chem. Soc. A*, 264.
Cottrell, T. L. (1954). "The Strengths of Chemical Bonds." Butterworths, London.
Coutts, R. S. P., and Wailes, P. C. (1968). *Chem. Commun.*, 260.
Coyle, T. D., Stafford, S. L., and Stone, F. G. A. (1961). *Spectrochim. Acta*, **17**, 968.
Cradock, S. (1968). *J. chem. Soc. A*, 1426.
Cradock, S., and Ebsworth, E. A. V. (1967). *J. chem. Soc. A*, 12, 1226.
Cradock, S., and Ebsworth, E. A. V. (1968). *J. chem. Soc. A*, 1420, 1423.
Cradock, S., Ebsworth, E. A. V., Davidson, G., and Woodward, L. A. (1967a). *J. chem. Soc. A*, 1229.
Cradock, S., Gibbon, G. A., and Van Dyke, C. H. (1967b). *Inorg. Chem.*, **6**, 1751.
Craig, D. P., Maccoll, A., Nyholm, R. S., Orgel, L. E., and Sutton, L. E. (1954). *J. chem. Soc.*, 332.
Cramer, R. D., Lindsey, R. V., Prewitt, C. T., and Stolberg, U. G. (1965). *J. Am. chem. Soc.*, **87**, 658.
Crawford, V. A., Rhee, K. H., and Wilson, M. K. (1962). *J. chem. Phys.*, **37**, 2377.
Creemers, H. M. J. C., and Noltes, J. G. (1967). *J. organometal. Chem.*, **7**, 237.
Cross, R. J. (1967). *Organometal. Chem. Rev.*, **2**, 97.
Cross, R. J., and Glockling, F. (1964). *J. chem. Soc.*, 4125.
Cross, R. J., and Glockling, F. (1965a). *J. organometal. Chem.*, **3**, 146.
Cross, R. J., and Glockling, F. (1965b). *J. organometal. Chem.*, **3**, 253.
Cross, R. J., and Glockling, F. (1965c). *J. chem. Soc.*, 5422.
Cruickshank, D. W. J. (1961). *J. chem. Soc.*, 5486.
Curtis, M. D. (1967). *J. Am. chem. Soc.*, **89**, 4241.
Curtis, M. D., and Allred, A. L. (1965). *J. Am. chem. Soc.*, **87**, 2554.
Curtis, M. D., Lee, R. K., and Allred, A. L. (1967). *J. Am. chem. Soc.*, **89**, 5150.
Dannley, R. L., and Farrant, G. (1966). *J. Am. chem. Soc.*, **88**, 627.
Davidsohn, W. E., and Henry, M. C. (1966). *J. organometal. Chem.*, **5**, 29.
Davidsohn, W. E., and Henry, M. C. (1967). *Chem. Rev.*, **67**, 73.
Davidsohn, W. E., Hills, K., and Henry, M. C. (1965). *Organometal. Chem.*, **3**, 285.
Davidson, G., Woodward, L. A., Mackay, K. M., and Robinson, P. (1967). *Spectrochim. Acta*, **23A**, 2383.
Davis, A. G., and Hall, C. D. (1959). *J. chem. Soc.*, 3835.
Davydov, V. I., and Diev, N. P. (1957). *Zh. neorg. Khim. (Eng. Transl.)*, **2**, 31.
Dede, L., and Russ, W. (1928). *Chem. Ber.*, **61B**, 2460.
Delman, A. D., Stein, A. A., Simms, B. B., and Katzenstein, R. J. (1966). *J. Polym. Sci.*, **4**, 2307.
Dennis, L. M., and Hance, F. E. (1922). *J. Am. chem. Soc.*, **44**, 2854.
Dennis, L. M., and Hance, F. E. (1925). *J. Am. chem. Soc.*, **47**, 370.
Dennis, L. M., and Hulse, R. E. (1930). *J. Am. chem. Soc.*, **52**, 3553.
Dennis, L. M., and Hunter, H. L. (1929). *J. Am. chem. Soc.*, **51**, 1151.
Dennis, L. M., and Joseph, S. M. (1927). *J. phys. Chem., Ithaca*, **31**, 1716.

Dennis, L. M., and Judy, P. R. (1929). *J. Am. chem. Soc.*, **51**, 2321.

Dennis, L. M., and Patnode, W. I. (1930). *J. Am. chem. Soc.*, **52**, 2779.

Dennis, L. M., and Skow, N. A. (1930). *J. Am. chem. Soc.*, **52**, 2369.

Dennis, L. M., and Work, R. W. (1933). *J. Am. chem. Soc.*, **55**, 4486.

Dennis, L. M., Corey, R. B., and Moore, R. W. (1924). *J. Am. chem. Soc.*, **46**, 657.

Dennis, L. M., Orndorff, W. R., and Tabern, D. L. (1926). *J. phys. Chem., Ithaca*, **30**, 1049.

Dibeler, V. H. (1952). *J. Res. natn. Bur. Stand.*, **49**, 235.

Dilts, J. A., and Johnson, M. P. (1966). *Inorg. Chem.*, **5**, 2079.

Dolgii, I. E., Meshcheryakov, A. P., and Gaivoronskaya, G. K. (1963). *Izv. Akad. Nauk. SSSR (Eng. Transl.)*, 519, 1009.

Drago, R. S. (1960). *J. inorg. nucl. Chem.*, **15**, 237.

Drake, J. E., and Jolly, W. L. (1962a). *Chemy Ind.*, 1470.

Drake, J. E., and Jolly, W. L. (1962b). *J. chem. Soc.*, 2807.

Drake, J. E., and Jolly, W. L. (1963). *Inorg. Synth.*, **7**, 34.

Drake, J. E., and Riddle, C. (1968). *J. chem. Soc. A*, 1675.

Drenth, W., Janssen, M. J., Van der Kerk, G. J. M., and Veiegenthart, J. A. (1964). *J. organometal. Chem.*, **2**, 265.

Drozdova, T. V., Kravtsova, R. P., and Tobelko, K. I. (1962). *Izv. Akad. Nauk. SSSR (Eng. Transl.)*, 31.

Dub, M. (1967). "Organometallic Compounds", 2nd Edn., Vol. 2, Springer Verlag, Berlin.

Duffield, A. M., Djerassi, C., Mazerolles, P., Dubac, J., and Manuel, G. (1968). *J. organometal. Chem.*, **12**, 123.

Dupuis, T. (1960). *Recl. Trav. chim. Pays-Bas Belge*, **79**, 578.

Durig, J. R., and Sink, C. W. (1968). *Spectrochim. Acta*, **24A**, 575.

Durig, J. R., Sink, C. W., and Bush, S. F. (1966). *J. chem. Phys.*, **45**, 66.

Dutta, S. N., and Jeffrey, G. A. (1965). *Inorg. Chem.*, **4**, 1363.

Dzhurinskaya, N. G., Mironov, V. F., and Petrov, A. D. (1961). *Dokl. Akad. Nauk SSSR (Eng. Transl.)*, **138**, 574.

Eaborn, C., and Pande, K. C. (1960a). *J. chem. Soc.*, 1566.

Eaborn, C., and Pande, K. C. (1960b). *J. chem. Soc.*, 3200.

Eaborn, C., and Pande, K. C. (1961). *J. chem. Soc.*, 297, 5082.

Eaborn, C., and Varma, I. D. (1967). *J. organometal. Chem.*, **9**, 377.

Eaborn, C., and Walton, D. R. M. (1964). *J. organometal. Chem.*, **2**, 95.

Eaborn, C., Leyshon, K., and Pande, K. C. (1960). *J. chem. Soc.*, 3423.

Eaborn, C., Simpson, P., and Varma, I. D. (1966a). *J. chem. Soc. A*, 1133.

Eaborn, C., Skinner, G. A., and Walton, D. R. M. (1966b). *J. organometal Chem.*, **6**, 438.

Eaborn, C., Dutton, W. A., Glockling, F., and Hooton, K. A. (1967). *J. organometal. Chem.*, **9**, 175.

Eaton, D. R., and McClellan, W. R. (1967). *Inorg. Chem.*, **6**, 2134.

Ebsworth, E. A. V., and Robiette, A. G. (1964). *Spectrochim. Acta*, **20**, 1639.

Egorochkin, A. N., Khidekel, M. L., Razuvaev, M. L., Mironov, V. F., and Kravchenko, A. L. (1964a). *Izv. Akad. Nauk SSSR (Eng. Transl.)*, 1214.

Egorochkin, A. N., Khidekel, M. L., Ponomarenko, V. A., Zueva, G. Ya., and Razuvaev, G. A. (1964b). *Izv. Akad. Nauk SSSR (Eng. Transl.)*, 347.

Egorov, Yu. P., Kirie, G. G., Leites, L. A., Mironov, V. F., and Petrov, A. D. (1962). *Izv. Akad. Nauk SSSR (Eng. Transl.)*, 1793.

Eisch, J. J., and Foxton, M. W. (1968). *J. organometal. Chem.*, **11**, P 24.

Eméleus, H. J., and Gardner, E. R. (1938). *J. chem. Soc.*, 1900.

8

Eméleus, H. J., and Jellinek, H. H. G. (1944). *Trans. Faraday Soc.*, **40**, 93.
Eméleus, H. J., and Mackay, K. M. (1961). *J. chem. Soc.*, 2676.
Eméleus, H. J., and Woolf, A. A. (1950). *J. chem. Soc.*, 164.
Emel'yanova, L. I., Vinogradova, V. N., Makarova, L. G., and Nesmeyanov, A. N. (1962). *Izv. Akad. Nauk SSSR (Eng. Transl.)*, 45.
Esposito, J. N., Sutton, L. E., and Kenney, M. E. (1967). *Inorg. Chem.*, **6**, 1116.
Evans, D. F., and Richards, R. E. (1952). *J. chem. Soc.*, 1292.
Everest, D. A. (1953). *J. chem. Soc.*, 4117.
Everest, D. A., and Salmon, J. E. (1954). *J. chem. Soc.*, 2438.
Everest, D. A., and Salmon, J. E. (1955). *J. chem. Soc.*, 1444.
Fensham, P. J., Tamaru, K., Boudart, M., and Taylor, H. (1955). *J. phys. Chem., Ithaca*, **59**, 806.
Fenton, D. E., Massey, A. G., and Urch, D. S. (1966). *J. organometal. Chem.*, **6**, 352.
Ferguson, J. E., Grant, D. K., Hickford, R. H., and Wilkins, C. J. (1959). *J. chem. Soc.*, 99.
Findeiss, W., Davidsohn, W. E., and Henry, M. C. (1967). *J. organometal. Chem.*, **9**, 435.
Finholt, A. E., Bond, A. C., Wilzbach, K. E., and Schlesinger, H. I. (1947). *J. Am. Chem. Soc.*, **69**, 2692.
Fish, R. H., and Kuivila, H. G. (1966). *J. org. Chem.*, **31**, 2445.
Fischer, A. K., West, R. C., and Rochow, E. G. (1954). *J. Am. chem. Soc.*, **76**, 5878.
Flitcroft, N., Harborne, D. A., Paul, I., Tucker, P. M., and Stone, F. G. A. (1966). *J. chem. Soc. A*, 1130.
Flood, E. A. (1932). *J. Am. chem. Soc.*, **54**, 1663.
Flood, E. A. (1933). *J. Am. chem. Soc.*, **55**, 4935.
Florinskii, F. S. (1962). *Zh. obshch. Khim. (Eng. Transl.)*, **32**, 1430.
Foster, L. S. (1946). *Inorg. Synth.*, **2**, 102.
Foster, L. S. (1950). *Inorg. Synth.*, **3**, 63.
Foster, L. S., and Williston, A. F. (1946). *Inorg. Synth.*, **2**, 112.
Foster, L. S., Drenan, J. W., and Williston, A. F. (1946). *Inorg. Synth.*, **2**, 109.
Frazer, M. J., Gerrard, W., and Spillman, J. A. (1964). *J. inorg. nucl. Chem.*, **26**, 1471.
Freeman, D. E., Rhee, K. H., and Wilson, M. K. (1963). *J. chem. Phys.*, **39**, 2908.
Fuchs, R., and Gilman, H. (1957). *J. org. Chem.*, **22**, 1009.
Fuchs, R., and Gilman, H. (1958). *J. org. Chem.*, **23**, 911.
Fuchs, R., Moore, L. O., Miles, D., and Gilman, H. (1956). *J. org. Chem.*, **21**, 1113.
Gar, T. K., and Mironov, V. F. (1966). *Zh. obshch. Khim. (Eng. Transl.)*, **36**, 1706.
Gastinger, E. (1956). *Z. anorg. Chem.*, **285**, 103.
Gayer, K. H., and Zajicek, O. T. (1964a). *J. inorg. nucl. Chem.*, **26**, 2123.
Gayer, K. H., and Zajicek, O. T. (1964b). *J. inorg. nucl. Chem.*, **26**, 951.
Gebala, A. E., and Jones, M. M. (1967). *J. inorg. nucl. Chem.*, **29**, 2301.
Geddes, R. L., and Mack, E. (1930). *J. Am. chem. Soc.*, **52**, 4372.
George, M. V., Talukdar, P. B., Gerow, C. W., and Gilman, H. (1960). *J. Am. chem. Soc.*, **82**, 4562.
George, M. V., Talukdar, P. B., and Gilman, H. (1966). *J. organometal. Chem.*, **5**, 397.
Gibbon, G. A., Wang, J. T., and Van Dyke, C. H. (1967). *Inorg. Chem.*, **6**, 1989.
Gilman, H., and Cartledge, F. K. (1966). *J. organometal. Chem.*, **5**, 48.
Gilman, H., and Gerow, C. W. (1955a). *J. Am. chem. Soc.*, **77**, 4675.
Gilman, H., and Gerow, C. W. (1955b). *J. Am. chem. Soc.*, **77**, 5509.
Gilman, H., and Gerow, C. W. (1955c). *J. Am. chem. Soc.*, **77**, 5740.

Gilman, H., and Gerow, C. W. (1956a). *J. Am. chem. Soc.*, **78**, 5435.
Gilman, H., and Gerow, C. W. (1956b). *J. Am. chem. Soc.*, **78**, 5823.
Gilman, H., and Gerow, C. W. (1957a). *J. org. Chem.*, **22**, 334.
Gilman, H., and Gerow, C. W. (1957b). *J. Am. chem. Soc.*, **79**, 342.
Gilman, H., and Gerow, C. W. (1958). *J. org. Chem.*, **23**, 1582.
Gilman, H., and Leeper, R. W. (1951). *J. org. Chem.*, **16**, 466.
Gilman, H., and Melvin, H. W. (1949). *J. Am. chem. Soc.*, **71**, 4050.
Gilman, H., and Smith, C. L. (1967). *J. organometal. Chem.*, **8**, 245.
Gilman, H., and Zuech, E. A. (1960). *J. Am. chem. Soc.*, **82**, 2522.
Gilman, H., and Zuech, E. A. (1961). *J. org. Chem.*, **26**, 3035.
Gilman, H., Hughes, M. B., and Gerow, C. W. (1959). *J. org. Chem.*, **24**, 352.
Gilman, H., Marrs, O. L., Trepka, W. J., and Diehl, J. W. (1962a). *J. org. Chem.*, **27**, 1260.
Gilman, H., Gorsich, R. D., and Gaj, B. J. (1962b). *J. org. Chem.*, **27**, 1023.
Gilman, H., Cartledge, F. K., and Sim, S. Y. (1963). *J. organometal. Chem.*, **1**, 8.
Gilman, H., Cartledge, F. K., and Sim, S. Y. (1965). *J. organometal. Chem.*, **4**, 332.
Gilman, H., Atwell, W. H., and Cartledge, F. K. (1966). *Adv. organometal. Chem.*, **4**, 1.
Gladshtein, B. M., Rode, V. V., and Soborovskii, L. Z. (1959). *Zh. obshch. Khim.* (*Eng. Transl.*), **29**, 2120.
Glarum, S. N., and Kraus, C. A. (1950). *J. Am. chem. Soc.*, **72**, 5398.
Glockling, F. (1966). *Q. Rev. chem. Soc.*, **20**, 45.
Glockling, F., and Hooton, K. A. (1962a). *J. chem. Soc.*, 2658.
Glockling, F., and Hooton, K. A. (1962b). *J. chem. Soc.*, 3509.
Glockling, F., and Hooton, K. A. (1963). *J. chem. Soc.*, 1849.
Glockling, F., and Hooton, K. A. (1966). *Inorg. Synth.*, **8**, 31.
Glockling, F., and Hooton, K. A. (1967). *J. chem. Soc. A*, 1066.
Glockling, F., and Hooton, K. A. (1968). *J. chem. Soc., A*, 826.
Glockling, F., and Light, J. R. C. (1967). *J. chem. Soc. A*, 623.
Glockling, F., and Light, J. R. C. (1968). *J. chem. Soc. A*, 717.
Glockling, F., and Wilbey, M. (1968). *J. chem. Soc. A*, 2168.
Glockling F., Light, J. R. C. and Walker, J. (1968). *Chem. Commun.*, 1052.
Gokhale, S. D., Drake, J. E., and Jolly, W. L. (1965). *J. inorg. nucl. Chem.*, **27**, 1911.
Goldfarb, T. D. (1962). *J. chem. Phys.*, **37**, 642.
Goldfarb, T. D., and Sujishi, S. (1964). *J. Am. chem. Soc.*, **86**, 1679.
Goldfarb, T. D., and Zafonte, B. P. (1964). *J. chem. Phys.*, **41**, 3653.
Gorsich, R. D. (1962). *J. Am. chem. Soc.*, **84**, 2486.
Grant, D., and Van Wazer, J. R. (1965). *J. organometal. Chem.*, **4**, 229.
Griffiths, J. E. (1963a). *J. chem. Phys.*, **38**, 2879.
Griffiths, J. E. (1963b). *Inorg. Chem.*, **2**, 375.
Griffiths, J. E. (1964). *Spectrochim. Acta*, **20**, 1335.
Griffiths, J. E. (1967). *Can. J. Chem.*, **45**, 2639.
Griffiths, J. E., and Onyszchuk, M. (1961). *Can. J. Chem.*, **39**, 339.
Gross, P., Hayman, C., and Bingham, J. T. (1966). *Trans. Faraday Soc.*, **62**, 2388.
Gunn, S. R., and Green, L. G. (1961). *J. phys. Chem., Ithaca*, **65**, 779.
Gunn, S. R., and Green, L. G. (1964). *J. phys. Chem., Ithaca*, **68**, 946.
Gutt, R. (1964). *Helv. chim. Acta*, **47**, 2262.
Gutmann, V., and Meller, A. (1960). *Mh. Chem.*, **91**, 519.

Gverdtsiteli, I. M., Guntsadze, T. P., and Petrov, A. D. (1964). *Dokl. Akad. Nauk SSSR (Eng. Transl.)*, **157**, 711.
Hague, D. N., and Prince, R. H. (1964). *Chemy Ind.*, 1492.
Harrah, L. A., Ryan, M. T., and Tamborski, C. (1962). *Spectrochim. Acta*, **18**, 21.
Harrison, R. W., and Trotter, J. (1968). *J. chem. Soc. A*, 258.
Hedberg, K. (1955). *J. Am. chem. Soc.*, **77**, 6491.
Henry, M. C., and Davidsohn, W. E. (1962). *J. org. Chem.*, **27**, 2252.
Henry, M. C., and Davidsohn, W. E. (1963). *Can. J. Chem.*, **41**, 1276.
Henry, M. C., and Downey, M. F. (1961). *J. org. Chem.*, **26**, 2299.
Henry, M. C., and Noltes, J. G. (1960). *J. Am. chem. Soc.*, **82**, 558.
Hoard, J. L., and Vincent, W. B. (1942). *J. Am. chem. Soc.*, **64**, 1233.
Hobrock, B. G., and Kiser, R. W. (1962). *J. phys. Chem., Ithaca*, **66**, 155.
Hoffman, C. J., and Gutowsky, H. S. (1953). *Inorg. Synth.*, **4**, 147.
Hogness, T. R., and Johnson, W. C. (1932). *J. Am. chem. Soc.*, **54**, 3583.
Holness, H. (1948). *Analyt. chim. Acta*, **2**, 254.
Hooton, K. A., and Allred, A. L. (1965). *Inorg. Chem.*, **4**, 671.
Huang, H., and Hui, K. (1964). *J. organometal. Chem.*, **2**, 288.
Huber, F., Enders, M., and Kaiser, R. (1966). *Z. Naturf.*, **21**, 83.
Husk, G. R., and West, R. (1965). *J. Am. chem. Soc.*, **87**, 3993.
Ibekwe, S. D., and Newlands, M. J. (1965). *J. chem. Soc.*, 4608.
Ingri, N. (1963). *Acta chem. scand.*, **17**, 597.
Ingri, N., and Lundgren, G. (1963). *Acta chem. scand.*, **17**, 617.
Ishii, Y., Itoh, K., Nakamura, A., and Sakai, S. (1967). *Chem. Commun.*, 224.
Issleib, K., and Walther, B. (1967). *Angew. Chem. Int. Ed.*, **6**, 88.
Itoh, K., Sakai, S., and Ishii, J. (1967). *Chem. Commun.*, 36.
Jacobs, G. (1954). *C.r. hebd. Séanc. Akad. Sci. Paris*, **238**, 1825.
Jaffee, R. I., McMullen, E. W., and Gonser, B. W. (1946). *J. electrochem. Soc.*, **89**, 277.
Jetz, W., Simons, P. B., Thompson, J. A. J., and Graham, W. A. G. (1966). *Inorg. Chem.*, **5**, 2217.
Jezowska, T. B., Hanuza, J., and Wojciechowski, W. (1967). *Spectrochim. Acta*, **23A**, 2631.
Johnson, E. B., and Dennis, L. M. (1925). *J. Am. chem. Soc.*, **47**, 790.
Johnson, F., Gohlke, R. S., and Nasutavicus, W. A. (1965). *J. organometal. Chem.*, **3**, 233.
Johnson, M. P., Shriver, D. F., and Shriver, S. A. (1966). *J. Am. chem. Soc.*, **88**, 1588.
Johnson, O. H. (1951). *Chem. Rev.*, **48**, 259.
Johnson, O. H. (1952). *Chem. Rev.*, **51**, 431.
Johnson, O. H., and Harris, D. M. (1950). *J. Am. chem. Soc.*, **72**, 5564.
Johnson, O. H., and Jones, L. V. (1952). *J. org. Chem.*, **17**, 1172.
Johnson, O. H., and Nebergall, W. H. (1948). *J. Am. chem. Soc.*, **70**, 1706.
Johnson, O. H., and Nebergall, W. H. (1949). *J. Am. chem. Soc.*, **71**, 1720.
Johnson, O. H., and Schmall, E. A. (1958). *J. Am. chem. Soc.*, **80**, 2931.
Johnson, O. H., Nebergall, W. H., and Harris, D. M. (1957). *Inorg. Synth.*, **5**, 74.
Johnson, W. C. (1930). *J. Am. chem. Soc.*, **52**, 5160.
Johnson, W. C., and Ridgely, G. H. (1934). *J. Am. chem. Soc.*, **56**, 2395.
Johnson, W. C., and Sidwell, A. E. (1933). *J. Am. chem. Soc.*, **55**, 1884.
Johnson, W. C., Morey, G. H., and Kott, A. E. (1932). *J. Am. chem. Soc.*, **54**, 4278.
Jolly, W. L. (1961). *J. Am. chem. Soc.*, **83**, 335.
Jolly, W. L., and Latimer, W. M. (1952a). *J. Am. chem. Soc.*, **74**, 5754.

Jolly, W. L., and Latimer, W. M. (1952b). *J. Am. chem. Soc.*, **74**, 5757.

Jones, K., and Lappert, M. F. (1965). *J. organometal. Chem.*, **3**, 295.

Joyner, R. D., and Kenney, M. E. (1960). *J. Am. chem. Soc.*, **82**, 5790.

Joyner, R. D., Linck, R. G., Esposito, J. N., and Kenney, M. E. (1962). *J. inorg. nucl. Chem.*, **24**, 299.

Kadina, M. A., Zueva, G. Ya., and Kechina, A. G. (1966). *Izv. Akad. Nauk SSSR (Eng. Transl.)*, 2145.

Kazarinova, N. F., and Vasilera, N. L. (1958). *Zh. analit. Khim. (Eng. Transl.)*, **13**, 765.

Kettle, S. F. A. (1959). *J. chem. Soc.*, 2936.

Kilbourn, B. T., Blundell, T. L., and Powell, H. M. (1965). *Chem. Commun.*, 444.

King, R. B. (1963). *Inorg. Chem.*, **2**, 199.

Klanberg, F. (1963). *Z. Naturf.*, **18**, 845.

Kolesnikov, S. P., and Nefedov, O. M. (1965). *Angew. Chem., Int. Ed.*, **4**, 352.

Kolesnikov, H. S., Davydova, S. L., and Klimentova, N. V. (1961). *J. Polym. Sci.*, **52**, 55.

Korshak, V. V., Polyakova, A. M., Mironov, V. F., and Petrov, A. D. (1959). *Izv. Akad. Nauk SSSR (Eng. Transl.)*, 169.

Kramer, K. A. W., and Wright, A. N. (1963). *Chem. Ber.*, **96**, 1877.

Kraus, C. A., and Brown, C. L. (1930a). *J. Am. chem. Soc.*, **52**, 3690.

Kraus, C. A., and Brown, C. L. (1930b). *J. Am. chem. Soc.*, **52**, 4031.

Kraus, C. A., and Carney, E. S. (1934). *J. Am. chem. Soc.*, **56**, 765.

Kraus, C. A., and Flood, E. A. (1932). *J. Am. chem. Soc.*, **54**, 1635.

Kraus, C. A., and Foster, L. S. (1927). *J. Am. chem. Soc.*, **49**, 457.

Kraus, C. A., and Sherman, C. S. (1933). *J. Am. chem. Soc.*, **55**, 4694.

Kruglaya, O. A., Razuvaev, G. A., Semchikova, G. S., and Vyazankin, N. S. (1966). *Dokl. Akad. Nauk SSSR (Eng. Transl.)*, **166**, 99.

Kruglaya, O. A., Vyazankin, N. S., Razuvaev, G. A., and Mitrofanova, E. V. (1967). *Dokl. Akad. Nauk SSSR (Eng. Transl.)*, **173**, 834.

Kuehlein, K., Neumann, W. P., and Becker, H. P. (1967). *Angew. Chem., Int. Ed.*, **6**, 876.

Kummer, R., and Graham, W. A. G. (1968). *Inorg. Chem.*, **7**, 310, 523.

Kurnevich, G. I., and Shagisultanova, G. A. (1964). *Z. neorg. Khim. (Eng. Transl.)*, **9**, 1383.

Ladenbauer, I. M., Siama, O., and Hecht, F. (1955). *Mikrochim. Acta*, 118.

Langer, H. G., and Blut, A. H. (1966). *J. organometal. Chem.*, **5**, 288.

Lappert, M. F. (1962). *J. chem. Soc.*, 542.

Laubengayer, A. W., and Brandt, P. L. (1932a). *J. Am. chem. Soc.*, **54**, 549.

Laubengayer, A. W., and Brandt, P. L. (1932b). *J. Am. chem. Soc.*, **54**, 621.

Laurie, V. W. (1959). *J. chem. Phys.*, **30**, 1210.

Lavigne, A. A., Pike, R. M., Monier, D., and Tabit, C. T. (1967). *Recl. Trav. Chim. Pays-Bas Belge*, **86**, 746.

Leavitt, F. C., Manuel, T. A., Johnson, F., Matternas, L. U., and Lehman, D. S. (1960). *J. Am. chem. Soc.*, **82**, 5099.

Leites, L. A., Gar, T. K., and Mironov, V. F. (1964). *Dokl. Akad. Nauk SSSR (Eng. Transl.)*, **158**, 894.

Lengel, J. H., and Dibeler, V. H. (1952). *J. Am. chem. Soc.*, **74**, 2683.

Lesbre, M., and Mazerolles, P. (1958). *C.r. hebd. Séanc. Acad. Sci. Paris*, **246**, 1708.

Lesbre, M., and Satge, J. (1962). *C.r. hebd. Séanc. Acad. Sci. Paris*, **254**, 1453, 4051.

Lesbre, M., Mazerolles, P., and Manuel, G. (1962). *C.r. hebd. Séanc. Acad. Sci. Paris*, **255**, 544.

Lesbre, M., Mazerolles, P., and Manuel, G. (1963). *C.r. hebd. Séanc. Acad. Sci. Paris*, **257**, 2303.
Leusink, A. J., Noltes, J. G., Budding, H. A., and Van der Kerk, G. J. M. (1964). *Recl. Trav. Chim. Pays-Bas Belge*, **83**, 844.
Levin, I. W. (1965). *J. chem. Phys.*, **42**, 1244.
Lieser, K. H., Elias, H., and Kohlschuetter, H. W. (1961). *Z. anorg. Chem.*, **313**, 199.
Lindeman, L. P., and Wilson, M. K. (1954). *J. chem. Phys.*, **22**, 1723.
Lindeman, L. P., and Wilson, M. K. (1957). *Spectrochim. Acta*, **9**, 47.
Lindsey, R. V., Parshall, G. W., and Stolberg, U. G. (1966). *Inorg. Chem.*, **5**, 109.
Lippincott, E. R., and Tobin, M. C. (1953). *J. Am. chem. Soc.*, **75**, 4141.
Lippincott, E. R., Mercier, P., and Tobin, M. C. (1953). *J. phys. Chem.*, *Ithaca*, **57**. 939.
Lohmann, D. H. (1965). *J. organometal. Chem.*, **4**, 382.
Luijten, J. G. A., and Rijkens, F. (1964). *Recl. Trav. Chim. Pays-Bas Belge*, **83**, 857.
Luitjen, J. G. A., Rijkens, F., and Van der Kerk, G. J. M. (1965). *Adv. organometal. Chem.*, **3**, 397.
Mackay, K. M. (1966). "Hydrogen Compounds of the Elements". Spon, London.
Mackay, K. M., and Roebuck, P. J. (1964). *J. chem. Soc.*, 1195.
Mackay, K. M., and Watt, R. (1967). *Spectrochim. Acta*, **23A**, 2761.
Mackay, K. M., and Watt, R. (1968). *J. organometal. Chem.*, **14,** 123.
Mackay, K. M., Robinson, P., Spanier, E. J., and MacDiarmid, A. G. (1966). *J. inorg. nucl. Chem.*, **28**, 1377.
Mackay, K. M., Sowerby, D. B., and Young, W. C. (1968a). *Spectrochim. Acta*, **24A**, 611.
Mackay, K. M., George, R. D., Robinson, P., and Watt, R. (1968b). *J. chem. Soc. A*, 1920.
McKean, D. C., and Chalmers, A. A. (1967). *Spectrochim. Acta*, **23A**, 777.
Mackenzie, J. D. (1958). *J. chem. Phys.*, **29**, 605.
Macklen, E. D. (1959). *J. chem. Soc.*, 1989.
Maddox, M. L., Stafford, S. L., and Kaesz, H. D. (1965). *Adv. organometal. Chem.*, **3**, 1.
Manulkin, Z. M., Kuchkarev, A. B., and Sarankina, S. A. (1963). *Dokl. Akad. Nauk SSSR (Eng. Transl.)*, **149**, 211.
Marrot, J., Maire, J. C., and Cassan, J. (1965). *C.r. hebd. Séanc. Acad. Sci. Paris*, **260**, 3931.
Massey, A. G., Park, A. J., and Stone, F. G. A. (1963). *J. Am. chem. Soc.*, **85**, 2021.
Massol, M., and Satge, J. (1966). *Bull. Soc. chim. Fr.*, 2737.
Masson, J. C., Le Quan, M., and Cadiot, P. (1967). *Bull. Soc. chim. Fr.*, 777.
Mathis, R., Satge, J., and Mathis, F. (1962). *Spectrochim. Acta*, **18**, 1463.
Mathis, R., Constant, M., Satge, J., and Mathis, F. (1964a). *Spectrochim. Acta*, **20**, 515.
Mathis, R., Sergent, M. C., Mazerolles, P., and Mathis, F. (1964b). *Spectrochim. Acta*, **20**, 1407.
Mathur, S., and Mehrotra, R. C. (1967). *J. organometal. Chem.*, **7**, 227.
Matwiyoff, N. A., and Drago, R. S. (1965). *J. organometal. Chem.*, **3**, 393.
Mays, J. M., and Dailey, B. P. (1952). *J. chem. Phys.*, **20**, 1695.
Mazerolles, P. (1960). *Bull. Soc. chim. Fr.*, 856.
Mazerolles, P. (1961). *Bull. Soc. chim. Fr.*, 1911.
Mazerolles, P. (1962). *Bull. Soc. chim. Fr.*, 1907.
Mazerolles, P., and Dubac, J. (1967). *C.r. hebd. Séanc. Acad. Sci. Paris*, **265**, 403.

Mazerolles, P., and Lesbre, M. (1959). *C.r. hebd. Séanc. Acad. Sci. Paris*, **248**, 2018.

Mazerolles, P., and Manuel, G. (1966). *Bull. Soc. chim. Fr.*, 327.

Mazerolles, P., Dubac, J., and Lesbre, M. (1966). *J. organometal. Chem.*, **5**, 35.

Mazerolles, P., Dubac, J., and Lesbre, M. (1968). *J. organometal. Chem.*, **12**, 143.

Mehrotra, R. C., and Chandra, G. (1963). *J. chem. Soc.*, 2804.

Mehrotra, R. C., and Mathur, S. (1966). *J. organometal. Chem.*, **6**, 11.

Mehrotra, R. C., and Mathur, S. (1967). *J. organometal. Chem.*, **7**, 233.

Mehrotra, R. C., Gupta, V. D., and Sukhani, D. (1967a). *J. inorg. nucl. Chem.*, **29**, 83.

Mehrotra, R. C., Gupta, V. D., and Sukhani, D. (1967b). *J. organometal. Chem.*, **9**, 263.

Mendeleeff, D. I. (1871). *Justus Liebigs Annln Chem.*, **8**, 196.

Mendelsohn, J. C., Metras, F., Lahournère, J. C., and Valade, J. (1968). *J. organometal. Chem.*, **12**, 327.

Metlesics, W., and Zeiss, H. (1960). *J. Am. chem. Soc.*, **82**, 3321, 3324.

Mikhailov, B. M., Bubnov, Yu. N., and Kiselev, V. G. (1965). *Izv. Akad. Nauk SSSR (Eng. Transl.)*, 58.

Miller, F. A., and Carlson, G. L. (1961). *Spectrochim. Acta*, **17**, 977.

Miller, J. M. (1967). *J. chem. Soc. A*, 828.

Miller, J. M., and Onyszchuk, M. (1967). *J. chem. Soc. A*, 1132.

Milligan, J. G. and Kraus, C. A. (1950). *J. Am. chem. Soc.*, **72**, 5297.

Mironov, V. F., and Dzhurinskaya, N. G. (1963). *Izv. Akad. Nauk SSSR (Eng. Transl.)*, 66.

Mironov, V. F., and Fedotov, N. S. (1966). *Zh. obshch. Khim. (Eng. Transl.)*, **36**, 574.

Mironov, V. F., and Gar, T. K. (1964a). *Izv. Akad. Nauk SSSR (Eng. Transl.)*, 1790.

Mironov, V. F., and Gar, T. K. (1964b). *Izv. Akad. Nauk SSSR. (Eng. Transl.)*, 1420.

Mironov, V. F., and Gar, T. K. (1965a). *Izv. Akad. Nauk SSSR (Eng. Transl.)*, 740, 827.

Mironov, V. F., and Gar, T. K. (1965b). *Izv. Akad. Nauk SSSR (Eng. Transl.)*, 273.

Mironov, V. F., and Gar, T. K. (1966). *Izv. Akad. Nauk SSSR (Eng. Transl.)*, 453.

Mironov, V. F., and Kravchenko, A. L. (1963). *Izv. Akad. Nauk SSSR (Eng. Transl.)*, 1425.

Mironov, V. F., and Kravchenko, A. L. (1964). *Dokl. Akad. Nauk. SSSR (Eng. Transl.)* **158**, 949.

Mironov, V. F., and Kravchenko, A. L. (1965). *Izv. Akad. Nauk SSSR (Eng. Transl.)*, 988.

Mironov, V. F., Egorov, Yu. P., and Petrov, A. D. (1959). *Izv. Akad. Nauk SSSR (Eng. Transl.)*, 1351.

Mironov, V. F., Dzhurinskaya, N. G., and Petrov, A. D. (1961). *Izv. Akad. Nauk SSSR (Eng. Transl.)*, 1956.

Mironov, V. F., Dzhurinskaya, N. G., Gar, T. K., and Petrov, A. D. (1962a). *Izv. Akad. Nauk SSSR (Eng. Transl.)*, 425.

Mironov, V. F., Gar, T. K., and Leites, L. A. (1962b). *Izv. Akad. Nauk SSSR (Eng. Transl.)*, 1303.

Mironov, V. F., Kravchenko, A. L., and Petrov, A. D. (1964a). *Dokl. Akad. Nauk SSSR (Eng. Transl.)*, **155**, 314.

Mironov, V. F., Kravchenko, A. L., and Petrov, A. D. (1964b). *Izv. Akad. Nauk SSSR* (*Eng. Transl.*), 1122.
Mironov, V. F., Kravchenko, A. L., and Leites, L. A. (1966). *Izv. Akad. Nauk SSSR* (*Eng. Transl.*), 1133.
Modern, E., and Wittman, A. (1965). *Mh. Chem.*, 96, 1783.
Moedritzer, K. (1966a). *Organometal. Chem. Rev.*, 1, 179.
Moedritzer, K. (1966b). *J. organometal. Chem.* 5, 254.
Moedritzer, K. (1967). *Inorg. Chem.*, 6, 1248.
Moedritzer, K., and Van Wazer, J. R. (1965a). *J. Am. chem. Soc.*, 87, 2360.
Moedritzer, K., and Van Wazer, J. R. (1965b). *Inorg. Chem.*, 4, 1753.
Moedritzer, K., and Van Wazer, J. R. (1967). *Inorg. chim. Acta*, 1, 152.
Moeller, T., and Nielsen, N. (1953). *J. Am. chem. Soc.*, 75, 5106.
Monnier, R., and Tissot, P. (1964). *Helv. chim. Acta*, 47, 2203.
Morgan, G., and Davies, G. R. (1937). *Chemy Ind.*, 56, 717.
Morgan, G. T., and Drew, H. D. K. (1924). *J. chem. Soc.*, 1261.
Morgan, G. T., and Drew, H. D. K. (1925). *J. chem. Soc.*, 1760.
Moulton, C. W., and Miller, J. G. (1956). *J. Am. chem. Soc.*, 78, 2702.
Mueller, J. H. (1922). *J. Am. chem. Soc.*, 44, 2493.
Mueller, J. H. (1926). *Proc. Am. Phil. Assoc.*, 65, 183.
Mueller, J. H., and Blank, H. R. (1924). *J. Am. chem. Soc.*, 46, 2358.
Mueller, J. H., Pike, E. F., and Graham, A. K. (1926). *Proc. Am. phil. Soc.*, 65, 15.
Mueller, R., and Heinrich, L. (1962). *Chem. Ber.*, 95, 2276.
Muetterties, E. L. (1960). *J. Am. chem. Soc.*, 82, 1082.
Muetterties, E. L. (1962). *Inorg. Chem.*, 1, 342.
Muetterties, E. L., and Castle, J. E. (1961). *J. inorg. nucl. Chem.*, 18, 148.
Murthy, M. K., and Hill, H. (1965). *J. Am. Ceram. Soc.*, 48, 109.
Musker, W. K., and Savitsky, G. B. (1967). *J. phys. Chem., Ithaca*, 71, 431.
Nagarajan, G. (1964). *Bull. Soc. chim. Belges*, 73, 874.
Nagy, J., Reffy, J., Borbely, A. K., and Becker, K. P. (1967). *J. organometal. Chem.*, 7, 393.
Nefedov, O. M., and Kolesnikov, S. P. (1966). *Izv. Akad. Nauk SSSR* (*Eng. Transl.*), 187.
Nefedov, O. M., and Manakov, M. N. (1963). *Izv. Akad. Nauk SSSR* (*Eng. Transl.*), 695.
Nefedov, O. M., and Manakov, M. N. (1966). *Angew. Chem., Int. Ed.*, 5, 1021.
Nefedov, O. M., Manakov, M. N., and Petrov, A. D. (1962). *Dokl. Akad. Nauk SSSR* (*Eng. Transl.*), 147, 1109.
Nefedov, O. M., Kolesnikov, S. P., and Novitskaya, N. N. (1965a). *Izv. Akad. Nauk SSSR* (*Eng. Transl.*), 568.
Nefedov, O. M., Szekely, T., Garzo, G., Kolesnikov, S. P., Manakov, M. N., and Shiryalv, V. I. (1965b). *Int. Symp. Organosilicon Chem.*, 65; *Chem. Abstr.*, 1966, 65, 12298.
Nefedov, O. M., Kolesnikov, S. P., and Perlmutter, B. L. (1967). *Angew. Chem., Int. Ed.*, 6, 628.
Nesmeyanov, A. N., Emel'yanova, L. I., and Makarova, L. G. (1958). *Dokl. Akad. Nauk SSSR* (*Eng. Transl.*), 122, 701.
Nesmeyanov, A. N., Anisimov, K. N., Kolobova, N. E., and Antonova, A. B. (1965a). *Izv. Akad. Nauk SSSR* (*Eng. Transl.*), 1284.
Nesmeyanov, A. N., Borisov, A. E., and Novikova, N. U. (1965b). *Dokl. Akad. Nauk SSSR* (*Eng. Transl.*), 165, 1090.

Nesmeyanov, A. N., Anisimov, K. N., Kolobova, N. E., and Antonova, A. B. (1966a). *Izv. Akad. Nauk SSSR (Eng. Transl.)*, 139, 142.

Nesmeyanov, A. N., Anisimov, K. N., Kolobova, N. E., and Denisova, F. S. (1966b). *Izv. Akad. Nauk SSSR (Eng. Transl.)*, 2185.

Neumann, W. P., and Kuehlein, K. (1965). *Justus Liebigs Annln Chem.*, **683**, 1.

Neumann, W. P., and Kuehlein, K. (1966). *Tetrahedron Lett.*, **29**, 3419.

Neumann, W. P., and Kuehlein, K. (1967a). *Justus Liebigs Annln Chem.*, **702**, 13.

Neumann, W. P., and Kuehlein, K. (1967b). *Justus Liebigs Annln Chem.*, **702**, 17.

Neumann, W. P., Schneider, B., and Sommer, R. (1966). *Justus Liebigs Annln Chem.*, **692**, 1.

Newlands, J. A. R. (1864). *Chem. News*, **10**, 59.

Noltes, J. G., and Van der Kerk, G. J. M. (1961). *Recl. Trav. Chim. Pays-Bas Belge*, **80**, 623.

Noltes, J. G., Budding, H. A., and Van der Kerk, G. J. M. (1960). *Recl. Trav. Chim. Pays-Bas Belge*, **79**, 1076.

Normant, H. (1967). *Angew. Chem., Int. Ed.*, **6**, 1046.

Org, W. K., and Prince, R. H. (1965). *J. inorg. nucl. Chem.*, **27**, 1037.

Orndorff, W. R., Tabern, D. L., and Dennis, L. M. (1927). *J. Am. chem. Soc.*, **49**, 2512.

Osipov, O. A., Shelepina, V. L., and Shelepin, O. E. (1966). *Zh. obshch. Khim. (Eng. Transl.)*, **36**, 274.

Parshall, G. W. (1966). *J. Am. chem. Soc.*, **88**, 704.

Patil, H. R. H., and Graham, W. A. G. (1966). *Inorg. Chem.*, **5**, 1401.

Patmore, D. J., and Graham, W. A. G. (1966). *Inorg. Chem.*, **5**, 1405.

Patmore, D. J., and Graham, W. A. G. (1967). *Inorg. Chem.*, **6**, 981.

Perkins, P. G. (1967). *Chem. Commun.*, 268.

Petukhov, G. G., Svirezheva, S. S., and Druzhkov, O. N. (1966). *Zh. obshch. Khim. (Eng. Transl.)*, **36**, 929.

Pflugmacher, A. K., and Hirsch, A. (1968). *J. organometal. Chem.*, **12**, 349.

Phillips, C. S. G., and Timms, P. L. (1963). *Analyt. Chem.*, **35**, 505.

Pichet, P., and Benoit, R. L. (1967). *Inorg. Chem.*, **6**, 1505.

Pike, R. M., and Dewidar, A. M. (1964). *Recl. Trav. Chim. Pays-Bas Belge*, **83**, 119.

Ponomarenko, V. A., and Vzenkova, G. Ya. (1957). *Izv. Akad. Nauk SSSR (Eng. Transl.)*, 1020.

Ponomarenko, V. A., Vzenkova, G. Ya., and Egorov, Yu. P. (1958). *Dokl. Akad. Nauk SSSR (Eng. Transl.)*, **122**, 703.

Ponomarenko, V. A., Zueva, G. Ya., and Andreev, N. S. (1961). *Izv. Akad. Nauk SSSR (Eng. Transl.)*, 1639.

Pope, A. E., and Skinner, H. A. (1964). *Trans. Faraday Soc.*, **60**, 1404.

Poskozim, P. S. (1968). *J. organometal. Chem.*, **12**, 115.

Prewitt, C. T., and Young, H. S. (1965). *Science, N.Y.*, **149**, 535.

Prince, R. H., and Timms, R. E. (1967). *Inorg. chim. Acta*, **1**, 129.

Pritchard, H. O., and Skinner, H. A. (1954). *Chem. Rev.*, **55**, 745.

Pugh, W. (1929). *J. chem. Soc.*, 1994.

Pullen, K. E., and Cady, G. H. (1967). *Inorg. Chem.*, **6**, 1300.

Quane, D., and Bottei, R. S. (1963). *Chem. Rev.*, **63**, 403.

Quane, D., and Hunt, G. W. (1968) *J. organometal. Chem.*, **14**, p. 16.

Rabinovich, I. B., Tel'noi, V. I., Karyakin, N. V., and Razuvaev, G. A. (1963). *Dokl. Akad. Nauk SSSR (Eng. Transl.)*, **149**, 216.

Randall, E. W., and Zuckerman, J. J. (1966). *Chem. Commun.*, 732.

8*

Randall, E. W., Ellner, J. J., and Zuckerman, J. J. (1966). *J. Am. chem. Soc.*, **88**, 622.

Rieche, A., and Dahlmann, J. (1964). *Justus Liebigs Annln Chem.*, **675**, 19.

Reichle, W. T. (1964). *Inorg. Chem.*, **3**, 402.

Rijkens, F. (1960). "Organogermanium Compounds". Germanium Research Committee, Utrecht, Holland.

Rijkens, F., and Van der Kerk, G. J. M. (1964). *Recl. Trav. Chim. Pays-Bas Belge*, **83**, 723.

Rijkens, F., Janssen, M. J., and Van der Kerk, G. J. M. (1965). *Recl. Trav. Chim. Pays-Bas Belge*, **84**, 1597.

Rijkens, F., Janssen, M. J., Drenth, W., and Van der Kerk, G. J. M. (1964). *J. organometal. Chem.*, **2**, 347.

Rijkens, F., Bulten, E. J., Drenth, W., and Van der Kerk, G. J. M. (1966). *Recl. Trav. Chim. Pays-Bas Belge*, **85**, 1223.

Roberts, R. M. G. (1968). *J. organometal. Chem.*, **12**, 97.

Roberts, R. M. G., and Kaissi, F. E. (1968). *J. organometal. Chem.*, **12**, 79.

Rochow, E. G. (1947). *J. Am. chem. Soc.*, **69**, 1729.

Rochow, E. G., and Allred, A. L. (1955). *J. Am. chem. Soc.*, **77**, 4489.

Rochow, E. G., Didtchenko, R., and West, R. C. (1951). *J. Am. chem. Soc.*, **73**, 5486.

Royen, P., and Rocktaeschel, C. (1966). *Z. anorg. Chem.*, **346**, 279.

Ruff, J. K. (1967a). *Inorg. Chem.*, **6**, 1502.

Ruff, J. K. (1967b). *Inorg. Chem.*, **6**, 2080.

Ruidisch, I. S., and Mebert, B. J. (1968). *J. organometal. Chem.*, **11**, 77.

Ruidisch, I., and Schmidt, M. (1963a). *J. organometal. Chem.*, **1**, 160.

Ruidisch, I., and Schmidt, M. (1963b). *Chem. Ber.*, **96**, 1424.

Ruidisch, I., and Schmidt, M. (1963c). *Chem. Ber.*, **96**, 821.

Ruidisch, I., and Schmidt, M. (1963d). *Z. Naturf.*, **18**, 508.

Ruidisch, I., and Schmidt, M. (1964). *Angew. Chem., Int. Ed.*, **3**, 231

Rustad, D. S., and Jolly, W. L. (1967). *Inorg. Chem.*, **6**, 1986.

Rustad, D. S., and Jolly, W. L. (1968). *Inorg. Chem.*, **7**, 213.

Saalfeld, E., and Svec, J. (1963). *Inorg. Chem.*, **2**, 46, 50.

Sacher, R. E., Lemmon, D. H., and Miller, F. A. (1967). *Spectrochim. Acta*, **23A**, 1169.

Sakurai, H., Tominaga, K., Watanabe, T., and Kumada, M. (1966). *Tetrahedron Lett.*, **45**, 5493.

Sarankina, S. A., and Manulkin, Z. M. (1966). *Zh. obshch. Khim. (Eng. Transl.)*, **36**, 1314.

Sartori, P., and Weidenbruch, M. (1967). *Chem. Ber.*, **100**, 2049.

Satge, J. (1961). *Annls. Chim.*, **6**, 519.

Satge, J. (1964). *Bull. Soc. chim. Fr.*, 630.

Satge, J., and Baudet, M. (1966). *C.r. hebd. Séanc. Acad. Sci. Paris C*, **263**, 435.

Satge, J., and Couret, C. (1967). *C.r. hebd. Séanc. Acad. Sci. Paris C*, **264**, 2169.

Satge, J., and Lesbre, M. (1965). *Bull. Soc. chim. Fr.*, 2578.

Satge, J., and Massol, M. (1965). *C.r. hebd. Séanc. Acad. Sci. Paris*, **261**, 170.

Satge, J., and Riviere, P. (1966). *Bull. Soc. chim. Fr.*, 1773.

Satge, J., Lesbre, M., and Baudet, M. (1964). *C.r. hebd. Séanc. Acad. Sci. Paris*, **259**, 4733.

Satge, J., Massol, M., and Lesbre, M. (1966). *J. organometal. Chem.*, **5**, 241.

Schafer, H. (1961). *Z. analyt. Chem.*, **180**, 15.

Schaefer, R., and Klemm, W. (1961). *Z. anorg. Chem.*, **312**, 214.

Scherer, O. J., and Biller, D. (1967). *Angew. Chem., Int. Ed.*, **6**, 446.

Scherer, O. J., and Schmidt, M. (1964). *J. organometal. Chem.*, **1**, 490.
Schlemper, E., and Britton, D. (1966). *Inorg. Chem.*, **5**, 507, 511.
Schmidbaur, H. (1964). *Chem. Ber.*, **97**, 1639.
Schmidbaur, H., and Hussek, H. (1964). *J. organometal. Chem.*, **1**, 235.
Schmidbaur, H., and Ruidisch, I. (1964). *Inorg. Chem.*, **3**, 599.
Schmidbaur, H., and Schmidt, M. (1961). *Chem. Ber.*, **94**, 1138, 1349, 2137.
Schmidbaur, H., and Tronich, W. (1967). *Chem. Ber.*, **100**, 1032.
Schmidt, M., and Ruf, H. (1963). *J. inorg. nucl. Chem.*, **25**, 557.
Schmidt, M., and Ruidisch, I. (1961). *Z. anorg. Chem.*, **311**, 331.
Schmidt, M., and Ruidisch, I. (1962). *Chem. Ber.*, **95**, 1434.
Schmidt, M., and Schumann, H. (1963). *Z. anorg. Chem.*, **325**, 130.
Schmidt, M., Schmidbaur, H., and Ruidisch, I. (1961a). *Angew. Chem.*, **73**, 408.
Schmidt, M., Schmidbaur, H., and Ruidisch, I. (1961b). *Chem. Ber.*, **94**, 2451.
Schmitz, D. O., and Jansen, W. (1967). *Z. anorg. Chem.*, **349**, 189.
Schott, G., and Harzdorf, C. (1961). *Z. anorg. Chem.*, **307**, 105.
Schumann, H. (1967). *Z. anorg. Chem.*, **354**, 192.
Schumann, I., and Blass, H. (1966). *Z. Naturf.*, **21**, 1105.
Schumann, H., and Stelzer, O. (1967). *Angew. Chem., Int. Ed.*, **6**, 701.
Schumann, H., Thom, K. F., and Schmidt, M. (1964a). *J. organometal. Chem.*, **2**, 97.
Schumann, H., Thom, K. F., and Schmidt, M. (1964b). *J. organometal. Chem.*, **2**, 361.
Schumann, H., Thom, K. F., and Schmidt, M. (1965). *J. organometal. Chem.*, **4**, 22.
Schumann, H., Schwabe, P., and Schmidt, M. (1966). *Inorg. nucl. Chem. Lett.*, **2**, 309, 313.
Schumann, R. I., Lieb, V., and Jutzi, M. B. (1967). *Z. anorg. Chem.*, **355**, 64.
Schumb, W. C., and Breck, D. W. (1952). *J. Am. chem. Soc.*, **74**, 1754.
Schumb, W. C., and Smyth, D. M. (1955). *J. Am. chem. Soc.*, **77**, 2133, 3003.
Schwarz, R., and Baronetzky, E. (1954). *Z. anorg. Chem.*, **275**, 1.
Schwarz, R., and Heinrich, F. (1932). *Z. anorg. Chem.*, **205**, 43
Schwarz, R., and Huf, E. (1931). *Z. anorg. Chem.*, **203**, 188.
Schwarz, R., and Krauff, K. G. (1954). *Z. anorg. Chem.*, **275**, 193.
Schwarz, R., and Lewinsohn, M. (1930). *Chem. Ber.*, **63**, 783.
Schwarz, R., and Lewinsohn, M. (1931). *Chem. Ber.*, **64**, 2352.
Schwarz, R., and Reinhardt, W. (1932). *Chem. Ber.*, **65**, 1743.
Schwarz, R., and Schenk, P. W. (1930). *Chem. Ber.*, **63**, 296.
Schwarz, R., and Schmeisser, M. (1936). *Chem. Ber.*, **69**, 579.
Schwarz, R., Schenk, P. W., and Giese, H. (1931). *Chem. Ber.*, **64**, 362.
Sedgwick, T. O. (1965). *J. electrochem. Soc.*, **112**, 496.
Semlyen, J. A., Walker, G. R., Blofield, R. E., and Phillips, C. S. G. (1964). *J. chem. Soc.*, 4948.
Semlyen, J. A., Walker, G. R., and Phillips, C. S. G. (1965). *J. chem. Soc.*, 1197.
Seyferth, D. (1957). *J. Am. chem. Soc.*, **79**, 2738.
Seyferth, D., and Alleston, D. L. (1963). *Inorg. Chem.*, **2**, 418.
Seyferth, D., and Cohen, H. M. (1962). *Inorg. Chem.*, **1**, 913.
Seyferth, D., and Dertouzos, H. (1968). *J. organometal. Chem.*, **11**, 263.
Seyferth, D., and Hetflejs, J. (1968). *J. organometal. Chem.*, **11**, 253.
Seyferth, D., and Kahlen, N. (1960a). *J. org. Chem.*, **25**, 809.
Seyferth, D., and Kahlen, N. (1960b). *J. Am. chem. Soc.*, **82**, 1080.
Seyferth, D., and Vaughn, L. G. (1963). *J. organometal. Chem.*, **1**, 138.

Seyferth, D., and Wada, T. (1962). *Inorg. Chem.*, **1**, 78.

Seyferth, D., and Weiner, M. A. (1962). *J. Am. chem. Soc.*, **84**, 361.

Seyferth, D., Raab, G., and Grim, S. O. (1961). *J. org. Chem.*, **26**, 3034.

Seyferth, D., Hofmann, H. P., Burton, R., and Helling, J. F. (1962a). *Inorg. Chem.*, **1**, 227.

Seyferth, D., Wada, T., and Maciel, G. E. (1962b). *Inorg. Chem.*, **1**, 232.

Seyferth, D., Suzuki, R., and Vaughn, L. G. (1966). *J. Am. chem. Soc.*, **88**, 286.

Seyferth, D., Burlitch, J. M., Dertouzos, H., and Simmons, H. D. (1967a). *J. organometal. Chem.*, **7**, 405.

Seyferth, D., Cross, R. J., and Prokai, B. (1967b). *J. organometal. Chem.*, **7**, P 20.

Shackelford, J. M., Schmertzing, H. D., Heuther, C. H., and Podall, H. (1963). *J. org. Chem.*, **28**, 1700.

Sharanina, L. G., Zavgorodnii, V. S., and Petrov, A. A. (1966). *Zh. obshch. Khim.* (*Eng. Transl.*), **36**, 1168.

Shevchenko, F. D., and Kuzina, L. A. (1963). *Ukr. khim. Zh.*, **29**, 351.

Shriver, D. F. (1966). "Structure and Bonding", Vol. 1, p. 32. Springer Verlag, Berlin.

Shriver, D. F., and Jolly, W. L. (1958). *J. Am. chem. Soc.*, **80**, 6692.

Sijpesteijn, A., Rijkens, F., Van der Kerk, G. J. M., and Marten, A. (1964). *J. Microbiol. Serol.*, **30**, 113.

Simons, J. K. (1935). *J. Am. chem. Soc.*, **57**, 1299.

Simons, J. K., Wagner, E. C., and Mueller, J. H. (1933). *J. Am. chem. Soc.*, **55**, 3705.

Sladkov, A. M., and Luneva, L. K. (1966). *Zh. obshch. Khim.* (*Eng. Transl.*), **36**, 570.

Smith, F. B., and Kraus, C. A. (1952). *J. Am. chem. Soc.*, **74**, 1418.

Smith, G. S., and Isaacs, P. B. (1964). *Acta crystallogr.*, **17**, 842.

Smith, J. A. S., and Wilkins, E. J. (1966). *J. chem. Soc. A*, 1749.

Soffer, H., and De Vries, T. (1951). *J. Am. chem. Soc.*, **73**, 5817.

Sollott, G. P., and Peterson, W. R. (1967). *J. Am. chem. Soc.*, **89**, 6783.

Sommer, L. H. (1962). *Angew. Chem., Int. Ed.*, **1**, 143.

Sosin, S. L., Korshak, V. V., and Alekseeva, V. P. (1964). *Vysokomolek. Soedin.*, (*Eng. Transl.*), **6**, 910.

Spanier, E. J., and MacDiarmid, A. G. (1963). *Inorg. Chem.*, **2**, 215.

Spialter, L., Buell, G. R., and Harris, C. W. (1965). *J. org. Chem.*, **30**, 375.

Srivastava, T. N., and Onyszchuk, M. (1963). *Can. J. Chem.*, **41**, 1244.

Srivastava, T. N., and Tandon, S. K. (1967). *Z. anorg. Chem.*, **353**, 87.

Srivastava, T. N., Griffiths, J. E., and Onyszchuk, M. (1962). *Can. J. Chem.*, **40**, 739.

Stafford, S. L., and Stone, F. G. A. (1961). *Spectrochim. Acta*, **17**, 412.

Steingross, W., and Zeil, W. (1966). *J. organometal. Chem.*, **6**, 109, 464.

Sterling, C. (1967). *J. inorg. nucl. Chem.*, **29**, 1211.

Steygers, L. (1960). *Mining Wld.*, Seattle, **22**, 34.

Stone, F. G. A. (1962). "Hydrogen Compounds of the Group IV Elements". Prentice-Hall, Princeton, N.J.

Storr, R., Wright, A. N., and Winkler, C. A. (1962). *Can. J. Chem.*, **40**, 1296.

Straley, J. W., Tindal, C. H., and Nielsen, H. H. (1942). *Phys. Rev.*, **62**, 161.

Strunz, H., and Giglio, M. (1961). *Acta crystallogr.*, **14**, 205.

Sujishi, S., and Keith, J. N. (1958). *J. Am. chem. Soc.*, **80**, 4138.

Tabern, D. L., Orndorff, W. R., and Dennis, L. M. (1925). *J. Am. chem. Soc.*, **47**, 2039.

Tamaru, K., Boudart, M., and Taylor, H. (1955). *J. phys. Chem.*, Ithaca, **59**, 801.

Tamborski, C., Ford, F. E., Lehn, W. L., Moore, G. J., and Soloski, E. J. (1962). *J. org. Chem.*, **27**, 619.

Tamborski, C., Soloski, E. J., and Dec, S. M. (1965). *J. organometal. Chem.*, **4**, 446.

Tananaev, I. V., Dzurinskii, B. F., and Mikhailov, Yu. N. (1964). *Z. neorg. Chem. (Eng. Transl.)*, **9**, 852.

Tchakirian, A. (1939). *Annls. Chim. Paris*, **12**, 415.

Tchakirian, A., and Carpéni, G. (1948). *C.r. hebd. Séanc. Acad. Sci. Paris*, **226**, 1094.

Teal, G. K., and Kraus, C. A. (1950). *J. Am. chem. Soc.*, **72**, 4706.

Thayer, J. S., and Strommen, D. P. (1966). *J. organometal. Chem.*, **5**, 383.

Thayer, J. S., and West, R. (1964). *Inorg. Chem.*, **3**, 889.

Thayer, J. S., and West, R. (1965). *Inorg. Chem.*, **4**, 114.

Thayer, J. S., and West, R. (1967). *Adv. organometal. Chem.*, **5**, 170.

Thomas, J. S., and Pugh, W. (1931). *J. chem. Soc.*, 60.

Thomas, J. S., and Southwood, W. W. (1931). *J. chem. Soc.*, 2083.

Tobias, R. S., and Hutcheson, S. (1966). *J. organometal. Chem.*, **6**, 535.

Treichel, P. M., and Stone, F. G. A. (1964). *Adv. organometal. Chem.*, **1**, 143.

Trotter, J., Akhtar, M., and Bartlett, N. (1966). *J. chem. Soc. A*, 30.

Tsau, J., Matsouo, S., Clerc, P., and Benoit, R. (1967). *Bull. Soc. chim. Fr.*, 1039.

Tzalmona, A. (1963). *Molec. Phys.*, **7**, 497.

Ulbricht, K., and Chvalovsky, V. (1968). *J. organometal. Chem.*, **12**, 105.

Van den Berghe, E. V., Van der Vondel, D. F., and Van der Kelen, G. P. (1967). *Inorg. chim. Acta*, **1**, 97.

Van der Kerk, G. J. M., Rijkens, F. and Janssen, M. J. (1962). *Recl. Trav. chim. Pays-Bas Belge*, **81**, 764.

Van der Vondel, D. F., and Van der Kelen, G. P. (1965a). *Bull. Soc. chim. Belge.*, **74**, 453.

Van der Vondel, D. F., and Van der Kelen, G. P. (1965b). *Bull. Soc. chim. Belge.*, **74**, 467.

Varma, R., and Buckton, K. S. (1967). *J. chem. Phys.*, **46**, 1575.

Varma, R., and Cox, A. P. (1967). *J. chem. Phys.*, **46**, 2007.

Vaska, L. (1966). *Chem. Commun.*, 614.

Vol'pin, M. E., Dulova, V. G., and Kursanov, D. N. (1963). *Izv. Akad. Nauk SSSR (Eng. Transl.)*, 649.

Vol'pin, M. E., Dulova, V. G., Struchkov, Yu. T., Bokii, N. K., and Kursanov, D. N. (1967). *J. organometal. Chem.*, **8**, 87.

Vyazankin, N. S., Razuvaev, G. A., and Gladyshev, E. N. (1963). *Dokl. Akad. Nauk SSSR (Eng. Transl.)*, **151**, 653.

Vyazankin, N. S., Razuvaev, G. A., Gladyshev, E. N., and Gurikova, T. G. (1964a). *Dokl. Akad. Nauk SSSR (Eng. Transl.)*, **155**, 360.

Vyazankin, N. S., Gladyshev, E. N., Korneva, S. P., and Razuvaev, G. A. (1964b). *Zh. obshch. Khim. (Eng. Transl.)*, **34**, 1656.

Vyazankin, N. S., Razuvaev, G. A., and Brevnova, T. N. (1965). *Zh. obshch. Khim. (Eng. Transl.)*, **35**, 2024.

Vyazankin, N. S., Gladyshev, E. N., Razuvaev, G. A., and Korneva, S. P. (1966a). *Zh. obshch. Khim. (Eng. Transl.)*, **36**, 952.

Vyazankin, N. S., Gladyshev, E. N., Korneva, S. P., and Razuvaev, G. A. (1966b). *Zh. obshch. Khim. (Eng. Transl.)*, **36**, 2025.

Vyazankin, N. S., Razuvaev, G. A., Bychkov, V. T., and Zvezdin, V. L. (1966c). *Izv. Akad. Nauk SSSR (Eng. Transl.)*, 533.

Vyazankin, N. S., Bochkarev, M. N., and Sanina, L. P. (1966d). *Zh. obshch. Khim.* (*Eng. Transl.*), **36**, 1154, 1961.

Vyazankin, N. S., Razuvaev, G. A., Gladyshev, E. N., and Korneva, S. P. (1967). *J. organometal. Chem.*, **7**, 353.

Walrafen, G. E. (1965). *J. chem. Phys.*, **42**, 485.

Wang, J. T., and Van Dyke, C. H. (1968). *Inorg. Chem.*, **7**, 1319.

West, R. (1952). *J. Am. chem. Soc.*, **74**, 4363.

West, R. (1953). *J. Am. chem. Soc.*, **75**, 6080.

West, R., and Baney, R. H. (1960). *J. phys. Chem.*, *Ithaca*, **64**, 822.

West, R., and Carberry, E. (1967). *Third Internat. Symp. organometallic Chem.*, 254.

West, R., Hunt, H. R., and Whipple, R. O. (1954). *J. Am. chem. Soc.*, **76**, 310.

West, R., Baney, R. H., and Powell, D. L. (1960). *J. Am. chem. Soc.*, **82**, 6269.

Whiffen, D. H. (1956). *J. chem. Soc.*, 1350.

Wiberg, E., Stecher, O., Andrascheck, H. J., Kreuzbichler, L., and Staude, E. (1963). *Angew. Chem., Int. Ed.*, **2**, 507.

Wiberg, E., Amberger, E., and Cambensi, H. (1967). *Z. anorg. Chem.*, **351**, 164.

Wieber, M., and Frohning, C. D. (1967). *J. organometal. Chem.*, **8**, 459.

Wieber, M., and Schmidt, M. (1963a). *Z. Naturf.*, **18**, 847, 849.

Wieber, M., and Schmidt, M. (1963b). *J. organometal. Chem.*, **1**, 93.

Wieber, M., Frohning, C. D., and Schmidt, M. (1966). *J. organometal. Chem.*, **6**, 427.

Wieber, M., Frohning, C. D., and Schwarzmann, G. (1967). *Z. anorg. Chem.*, **355**, 79.

Wilkinson, G. R., and Wilson, M. K. (1966). *J. chem. Phys.*, **44**, 3867.

Willemsens, L. C., and Van der Kerk, G. J. M. (1964). *J. organometal. Chem.*, **2**, 260.

Wingleth, D. C., and Norman, A. D. (1967). *Chem. Commun.*, 1218.

Winkler, C. (1887). *J. prakt. Chem.*, **36**, 177.

Wittle, J. K., and Urry, G. (1968). *Inorg. Chem.*, **7**, 560.

Woltz, P. J. H., and Nielsen, A. H. (1952). *J. chem. Phys.*, **20**, 307.

Worrall, D. E. (1940). *J. Am. chem. Soc.*, **62**, 3267.

Wunderlich, E., and Gohring, E. (1959). *Z. analyt. Chem.*, **169**, 346.

Yoder, C. H., and Zuckerman, J. J. (1964). *Inorg. Chem.*, **3**, 1329.

Yoder, C. M. S., and Zuckerman, J. J. (1967). *Inorg. Chem.*, **6**, 163.

Young, J. F., Gillard, R. D., and Wilkinson, G. (1964). *J. chem. Soc.*, 5176.

Zachariasen, W. (1928). *Z. Kristallogr. Krittallgeom.*, **67**, 226.

Zakharkin, L. I., and Okhlobystin, O. Yu. (1961). *Zh. obshch. Khim.* (*Eng. Transl.*), **31**, 3417.

Zakharkin, L. I., Okhlobystin, O. Yu., and Strunin, B. N. (1962). *Izv. Akad. Nauk SSSR* (*Eng. Transl.*), 1913.

Zakharkin, L. I., Bregadze, V. I., and Okhlobystin, O. Yu. (1965). *J. organometal. Chem.*, **4**, 211.

Zeltmann, A. H., and Fitzgibbon, G. C. (1954). *J. Am. chem. Soc.*, **76**, 2021.

Zmbov, K. F., Hastie, J. W., Hauge, R., and Margrave, J. L. (1968). *Inorg. Chem.*, **7**, 608.

Zueva, G. Ya., Luk'yankina, N. V., Kechina, A. G., and Ponomarenko, V. A. (1966). *Izv. Akad. Nauk SSSR* (*Eng. Transl.*), 1780.

Zumbusch, M., Heimbrecht, M., and Biltz, W. (1948). *Z. anorg. Chem.*, **242**, 237.

AUTHOR INDEX

Numbers in *italics* refer to the page on which the reference is listed.

SUBJECT INDEX

A

Acceptor properties,
  of GeX$_2$, 34–35
  of GeX$_4$, 39, 41–45
  of R$_3$GeX, 94
Acetylacetonates of Ge, 44–45, 109
Acetylenes, addition,
  of Ge dihalides, 72–73, 139
  of Ge hydrides, 91–93, 134–138
Alcohols, condensation with GeH and
  GeX, 109–110, 137
Alkali metal germanides, 28
Alkali metal derivatives,
  of GeH$_4$ and Ge$_2$H$_6$, 49–50, 55
  of the type R$_2$GeM$_2$, 85, 90, 165–167
  of the type R$_3$GeM, 85–86, 89–90,
    149–150, 154–155, 158–167
Alkenes, addition of,
  GeX$_2$, 71–73
  of germanium hydrides, 91–93, 134–
    138
Alkenylgermanes, 65–68, 127–130, 134–
  138
Alkoxy-Ge compounds, 37–38, 54–56,
  108–110
Alkylation,
  of Ge–H, 85–93, 134–138
  of Ge–X, 58–78, 122–125, 138–140
Alkyl- and aryl-germanes,
  formation, 58–79
  reactions, 79–83
Alkynylgermanes, 68–69, 138
Alloys of Ge, 28–29, 46
Aluminium-alkyls and alkyl hydrides,
  60–61, 66, 149
Aluminium-Ge compounds, 29, 149,
  173
Aminogermanes, 42–45, 50, 54–57, 94–
  98, 142
Analytical methods, 29
Antimony-Ge compounds, 173
Appearance potentials, 10–12
Argyrodite, 1, 28
Arsenate of Ge, 38
Arylgermanes, 58–83

Asymmetric germanes, 63–65, 131
Azidogermanes, 55, 144–145

B

Barium-Ge compounds, 168
Basicities,
  of Ge–O compounds, 55–56, 103–
    104
  of Ge–S compounds, 55–56
Bismuth-Ge compounds, 174
Bonding in Ge compounds, 4–9, 54–56,
  77
Bond-energies, 10–12, 48
  lengths, 12–13
Borohydrides of Ge, 47, 55
Boron-Ge compounds, 173
Boron trifluoride, cleavage
  of Ge–O, 105
  of Ge–C, 129
Bromination,
  of Ge–H, 50–51, 105, 131–132
  of Ge–C, 125–130
  of Ge–Ge, 157
  of Ge–P, 99
Bromogermanes, 50–53, 122–141
Butadiene, addition of Ge–H, 136

C

Cadmium-Ge compounds, 171
Calcium germanide, 28
Carbene insertion,
  into Ge–H, 91
  into Ge–Hg, 172
Carboxylic acids and esters of organogermanes, 65, 75, 168–169
Catenated germanes, 46–49, 52, 141,
  147–158
Chlorogermanes, 32–33, 40–45, 50–57,
  122–145
Chromium-Ge compounds, 176
Cleavage of ethers,
  by HGeCl$_3$, 135
  by R$_3$GeLi, 164–166